Science

PEP Revision

Workbook

GRADE 6

Amani Leslie

Every effort has been made to trace all copyright holders, but if any have been inadvertently overlooked, the Publishers will be pleased to make the necessary arrangements at the first opportunity.

Although every effort has been made to ensure that website addresses are correct at time of going to press, Hodder Education cannot be held responsible for the content of any website mentioned in this book. It is sometimes possible to find a relocated web page by typing in the address of the home page for a website in the URL window of your browser.

Hachette UK's policy is to use papers that are natural, renewable and recyclable products and made from wood grown in well-managed forests and other controlled sources. The logging and manufacturing processes are expected to conform to the environmental regulations of the country of origin.

Orders: please contact Hachette UK Distribution, Hely Hutchinson Centre, Milton Road, Didcot, Oxfordshire, OX11 7HH. Telephone: +44 (0)1235 827827. Email: education@hachette.co.uk. Lines are open from 9 a.m. to 5 p.m., Monday to Friday. You can also order through our website: www.hoddereducation.com.

ISBN: 978 1398313 330

© Hodder & Stoughton Limited 2021

First published in 2021 by

Hodder Education,

An Hachette UK Company

Carmelite House

50 Victoria Embankment

London EC4Y 0DZ

www.hoddereducation.com

Impression number   10 9 8 7 6 5 4 3 2

Year   2024 2023

All rights reserved. Apart from any use permitted under UK copyright law, no part of this publication may be reproduced or transmitted in any form or by any means, electronic or mechanical, including photocopying and recording, or held within any information storage and retrieval system, without permission in writing from the publisher or under licence from the Copyright Licensing Agency Limited. Further details of such licences (for reprographic reproduction) may be obtained from the Copyright Licensing Agency Limited, www.cla.co.uk.

Cover illustration by Heather Clarke c/o D'Avila Illustration

Illustrations by James Hearne, Natalie and Tamsin Hinrichsen, Val Myburgh, Vian Oelofsen and Hyphen S.A.

Typeset in FS Albert 15/17 by Hyphen S.A.

Printed and bound by CPI Group (UK) Ltd, Croydon, CR0 4YY

A catalogue record for this title is available from the British Library.

Contents

Term 1 Unit 1: The environment

The environment	5
Different types of environment	9
Types of soil	12
Soil investigation	14
Animal adaptations	17
Plant adaptations	21
Types of environmental change	24
How humans damage the environment	27
Conservation	30
Self-check	33
Climate change	35
Effects of climate change	38
The greenhouse effect	40
Causes and effects of soil degradation	43
Solid waste pollution	47
Reducing waste	49
Self-check	52
Extension activity	53
Practice test	55

Term 1 Unit 2: Light and sound

Luminous or non-luminous?	59
How does light travel?	63
Translucent, transparent, opaque	66
Dull or shiny?	68
Reflection 1	70
Reflection 2	71
Refraction	73
Lenses	75
Self-check	78
What is sound?	79

Pitch	82
Speed of sound	85
Volume	87
Noise pollution	90
Self-check	93
Extension activity	94
Practice test	97

Term 2 Unit 1: Materials and their properties

Different types of materials	102
Properties of different materials	105
Investigating properties	107
Relating properties to uses	109
Disposing of materials	113
Recycling	115
Self-check	118
Solids, liquids and gases	119
Properties of solids, liquids and gases	121
Changes of state	123
Heating and cooling	126
Irreversible and reversible changes	128
Examples of irreversible and reversible changes	131
Self-check	134
Extension activity	135
Practice test	138

Term 2 Unit 2: Human body systems

Different body systems	143
The circulatory system	146
The respiratory system	148
The male reproductive system	151
The female reproductive system	152

Contents

The digestive system 1 153
The digestive system 2 156
The skeletal system 159
Muscles and joints 162
Investigating movement 165
Excretion 169
The nervous system 172
Self-check 175
Extension activity 176
Practice test 179

Term 3 Unit 1: Mixtures

What is a mixture? 184
Classifying mixtures 187
Separating mixtures – filtration
and sieving 189
Separating mixtures – decanting and
evaporation 192
Separating using magnetism 195
Investigating separating mixtures 198
Self-check 201
Extension activity 202
Practice test 205

Term 3 Unit 2: Diet and drugs

Balanced diet 209
Obesity 210
Diabetes 212
Malnutrition 215
Deficiency diseases 218
Self-check 221
Legal and illegal drugs 222
More about illegal drugs 225
Medicinal drugs 228
Alcohol 231
Tobacco 234
Lifestyle and health 237

Self-check 240
Extension activity 241
Practice test 244

Acknowledgements 248

Term 1 Unit 1 | The environment

The environment

Learning objectives

- Formulate a definition of the environment.
- Describe different factors in different environments as living and non-living.
- Describe some ways to measure the different factors in environments.

Term 1 Unit 1 The environment

Look at these two pictures.

Can you describe the conditions in the two different places?

Think about the temperature of each place, if it is light or shaded and if it is dry or rainy.

....

....

....

....

....

....

....

....

....

....

In this lesson, you will learn about what an environment is. You will look at the conditions in different environments and describe some ways to measure these differences.

The environment

What is an environment?

1. Read these children's definitions of the environment.

   > The environment is the place where we live.

   > The environment is all external factors that affect us.

   Which do you think is the best definition?

2. Actually, these definitions are both correct. Do you think you can write a definition that includes both of these statements?

   Write your definition in the space below.

   ....

   ....

> **Make a note**
>
> Different factors in the environment can be split into two groups: living factors (biotic) and non-living factors (abiotic).

What are the different factors?

1. Look at the list of factors below. Separate them into living and non-living factors and write them in the correct place in the table.

   One living and one non-living factor have been done for you.

   disease   food   light   oxygen   ~~parasites~~   soil
   temperature   water   ~~wind~~

Living factors	Non-living factors
parasites	wind

7

Term 1 Unit 1 The environment

2 You can measure the differences in the factors in the environment using different methods.

Match the factor with the way that we can measure it by joining the boxes.

Factor
light
parasites
soil
temperature

Way to measure
population size
particle size
thermometer
light meter

Quick quiz

Are these statements **true** (**T**) or **false** (**F**)?

1 The environment includes the conditions we live in.   T   F
2 Number of predators in the environment is a non-living
   (abiotic) factor.   T   F
3 I can measure the rainfall with a measuring cylinder.   T   F

Different types of environment

> **Learning objectives**
> - Describe different environments.
> - Investigate features of different environments.
> - Evaluate information to place animals and plants in their natural environment.

Think about the conditions in your environment right now. Can you describe them? Record as much information as you can in the space below. If it is possible, see if you can measure some of the factors such as temperature.

....

....

....

In this lesson, you will look at the conditions in your own and different environments. You will be able to describe the conditions in different environments and match animals and plants to their environment.

Different types of environment

1. Read these children's descriptions of four different environments.

> Desert – This place is very hot and sandy. There is no rain or shade. An example is the Sahara desert.

> Marine – This place is very salty and wet. It includes oceans, seas, bays and estuaries. An example is the Caribbean Sea.

> Tundra – This place is very cold and is covered with snow and ice. An example is the Arctic.

> Rainforest – This place is covered in trees. It is very warm and wet. An example is the Brazilian rainforest.

Term 1 Unit 1 The environment

2 Think about what humans would need to visit these environments. For each of the options below, say what equipment you think you will need. Think about what would be appropriate clothing. Are you able to breathe in this environment? Do you need shade or water?

Type of environment	Equipment needed
desert	
marine	
tundra	
rainforest	

Different animals and plants in these environments

The boxes show some different animals and plants.

Can you match them to the environment they live in by linking the correct boxes? Think about what features they have and what they need to survive in each environment.

	camel		marine
	polar bear		rainforest
	shark		desert
	kapok tree		tundra

Different types of environment

The environment in which you live

Can you write lists of animals and plants that live in your environment?

Animals	Plants

Quick quiz

1 Name a freshwater environment local to you. ....
2 Name a marine environment local to you.   ....
3 What are the conditions like in a desert?   ....
4 Name a cold and icy environment.   ....

Term 1 Unit 1 The environment

Types of soil

> **Learning objectives**
> - Describe soil of my home environment.
> - Investigate soils of different environments.
> - Analyse particle size of soil.

Collect a sample of soil from your garden or an area local to you.

Take a photograph of it or draw a picture of it.

Write a description of it. Think about what colour it is, how it feels (texture), whether it is damp or dry and if the particles are fine and very small or big.

> ⚠ **Safety**
> Remember to wash your hands after handling soil.

....
....
....
....
....
....

In this lesson, you will look at soil in different environments. You will be able to describe the soil of your home environment, and analyse the particle size of soil in your environment and other environments.

Types of soil

Particle size experiment

1. Follow these instructions to have a more detailed look at your soil sample.
   a) Collect sieves with different size holes (ask an adult first).
   b) Use the sieve with the largest holes and put your soil sample in it.
   c) Shake the sieve and collect anything that comes out of the sieve.
   d) The soil particles left in the sieve will be the largest particles. Put these on to a piece of paper.
   e) Repeat this process for all the sizes of holes in the sieves that you have using the next smallest size each time.
   f) Put the particles you have collected in order of size from smallest to largest.

2. Answer these questions based on your experiment.
   a) How many different size particles did you collect? ....
   b) Which particle size did you collect the most of? ....
   c) Measure the width of the largest particle size with a ruler and write it down. Include the unit. ....

Different types of soil

Different types of soil will have different properties. Think about the soil on a beach and in a wetland environment. If possible, see if you can collect samples of these soils.

Circle the words that you think would best describe the soil from these three environments.

Type of environment	Words to choose from						
beach (at low tide)	crumbly	dry	fine	hard	large	soft	wet
dry forest	crumbly	dry	fine	hard	large	soft	wet
wetland	crumbly	dry	fine	hard	large	soft	wet

Term 1 Unit 1 The environment

Soil investigation

> **Learning objectives**
> - Plan an investigation to test the water-holding capacity of different types of soil.
> - Write a method identifying the dependent, independent and control variables.

Think about the things you need to do when planning an investigation.

Tick (✓) the things you need to do.

writing a list of apparatus		collecting results		writing a conclusion	
drawing a graph		writing a method		writing a hypothesis	
identifying variables		suggesting improvements			

In this lesson, you will plan an investigation to test the water-holding capacity of different soils.

Planning an investigation

Read the information and use it to help you to fill out the planning sheet on the next page.

You are investigating how much water each soil sample holds.

You will put each soil sample in a funnel with some cotton wool in the bottom and add some water.

You will measure how much water comes out over a set amount of time.

Questions to think about:

How will you measure the volume of water? How long will you time it for? How much soil will you use? What are you trying to find out? What variables will you keep the same in your investigation?

14

Soil investigation

You will need:

- soil samples
- funnels
- beakers
- measuring cylinders
- sieves
- cotton wool
- water
- a ruler
- a stop clock.

Writing your plan

1 Write a hypothesis for your investigation.

....

....

Hint

What are you trying to find out?

2 Write down the dependent variable for this investigation.

....

....

Hint

What are you going to measure?

3 Write down the independent variable for this investigation.

....

....

Hint

What are you going to change?

4 Write down which variable you will control (keep the same).

....

5 Write down your apparatus and a step-by-step method in the table on the next page.

Term 1 Unit 1 The environment

List of apparatus you will use	Step-by-step method
	a) ....
	b) ....
	c) ....
	d) ....
	e) ....
	f) ....

⚠️ Safety

Write down things you should do during the investigation to keep you safe.

6 ....
....

Quick quiz

Are these statements **true** (T) or **false** (F)?

1. The independent variable is the variable you measure.   T   F
2. It is important to control variables to make the investigation a fair test.   T   F
3. You should always wash your hands after handling soil.   T   F

Animal adaptations

> **Learning objectives**
> - Identify some special features that enable animals to live in different environments.
> - Research the adaptations of one Jamaican animal.
> - Describe different animal adaptations in relation to their environment.

Look at the environment in the photo.

What special features do you think you would need to be able to survive and find food in this environment?

Write down your ideas.

....

....

....

....

In this lesson, you will find out about different adaptations that animals have to be able to survive in different environments. You will be able to suggest the special adaptations an animal has and match it with its environment.

Term 1 Unit 1 The environment

What special features does an animal have?

ICT opportunity

You are going to research an animal that you find in Jamaica and find out what special features it has in order to live in its environment.

You can choose one of the following organisms or your own:

    Jamaican boa   manatee   crocodile   sea turtle

1. Research your animal on the internet or in books.
2. Answer the questions about the animal that you have researched.

    a) Draw a picture of your animal.

Animal adaptations

b) What is your animal's scientific name?

....

c) What is the environment that your animal lives in?

....

....

d) What are the conditions like in that environment?

....

....

....

....

e) What special features does your animal have? Think about how they move / how they get their food / how they hide from predators / how they attract mates.

....

....

....

....

....

....

....

....

Term 1 Unit 1 The environment

Adaptations of different animals

Make a note

Adaptations are special features an animal has. Read the information about the special features some different animals have.

Can you match the animal to the feature and the environment?

Animal
camel
lion
seal

Feature
thick layer of blubber to prevent getting cold
long eyelashes to get sand out of eyes
sandy fur for camouflage

Environment
grassland
Arctic sea
desert

Quick quiz

1 Name one feature that an animal that lives in a cold environment has. ....

2 Name one feature that an animal that lives in a hot environment has. ....

3 Name one feature that an animal that lives in the Arctic sea has. ....

Plant adaptations

> **Learning objectives**
> - Identify some special features that enable plants to live in different environments.
> - Research the adaptations of one Jamaican plant.
> - Describe different plant adaptations in relation to their environment.

Think about some of the different plants in your local environment.

1 Write down the names of three plants you know.

   a) ....   b) ....   c) ....

2 Describe one thing that makes each plant look different from the other two plants.

   a) ....

   b) ....

   c) ....

In this lesson, you will find out about different adaptations that plants have in order to be able to survive in different environments. You will be able to suggest the special adaptations a plant has and match it with its environment.

3 Match the adaptation with the benefit for the plant by joining the boxes.

Adaptation
long roots
thorns
large flowers

Benefit
to stop predators eating it
to attract insects for pollination
to get better access to water

Term 1 Unit 1 The environment

What special features does a plant have?

ICT opportunity

You are going to research a plant that you find in Jamaica and find out what special features it has in order to live in its environment.

You can choose one of the following organisms or your own:

Jamaican mangrove tree   aloe   maiden plum

1 Research your plant on the internet or in books.

2 Answer the questions about the plant that you have researched.

   a) Draw a picture of your plant.

   b) What is your plant's scientific name?

   ....

   c) What is the environment that your plant lives in?

   ....

   d) What are the conditions like in that environment?

   ....

   ....

   ....

Plant adaptations

e) What special features does your plant have? Think about how they get light / how they protect themselves from predators / how they survive the environment they live in / how they attract insects for pollination.

....

....

....

Adaptations of a cactus

Look at the photograph of a plant.

1 What environment do you think this plant lives in?

   ....

   ....

2 How do you think the following features of the plant help it to live in its environment?

   a) spikes instead of leaves

      ....

   b) thick fleshy stem

      ....

   c) thick waterproof outer layer

      ....

Quick quiz

Are these statements **true** (T) or **false** (F)?
1 Thorns can protect against predators.   T   F
2 Large leaves prevent the plant from losing water.   T   F
3 Only desert plants need special adaptations.   T   F

Term 1 Unit 1 The environment

Types of environmental change

> **Learning objectives**
> - Describe different environmental changes that occur and their impacts.
> - Recognise the need for and importance of conserving living things and the environment to sustain the balance in the ecosystem.

Think about the changes that happen in the environment. Many of them have an impact on the humans, animals and plants living in that environment.

Can you name some natural disasters? Write down as many as you can think of.

....

....

....

....

....

....

....

....

....

....

In this lesson, you will look at a case study of an environmental change and examine its impacts.

Case study: 1986 Jamaican floods

1 Read this newspaper article about the 1986 floods in Jamaica.

The Jamaican Daily

$1 June 8th 1986

Storm Andrew causes devastating floods in Jamaica

Torrential rainfall began falling in Jamaica on 24 May and has continued for the last two weeks. Some places have received as much as 25 inches of rain. This has caused widespread flooding and landslides and an estimated total of $25 million of damage.

Transport links were severely disrupted with 300 roads and 15 bridges damaged. The human cost is extreme with 50 people having lost their lives and 2,000 people homeless.

Agriculture and water supplies have been affected. Crops have been destroyed and 100,000 people are without water.

The salinity of seawater has decreased and so has the temperature of the seawater due to freshwater flooding. This has had an adverse effect on many of the fish species that live off the coast of Jamaica.

2 Use the information in the newspaper article and answer the following questions.

 a) Name the two environmental changes that the heavy rainfall caused.

 ....

 ....

 b) What was the overall estimated cost of the damage caused by the 1986 flooding?

 ....

Term 1 Unit 1 The environment

    c) Describe how humans were affected by these environmental changes.

    ....

    ....

    ....

    ....

    d) Describe how animals and plants were affected by these environmental changes.

    ....

    ....

    ....

3 Think about the environmental damage to crops and animals, such as fish, that is caused by flooding.

Suggest why it is important that we try to conserve these things in particular.

Hint

Think about food sources, economic value and biodiversity.

....

....

....

....

Quick quiz

1 What is a drought? ....
2 How does drought affect crops? ....
3 What is a hurricane? ....

How humans damage the environment

Learning objectives
- Describe some activities by humans that damage the environment.
- Outline the effects of human activities on the environment.
- Recognise the need for and importance of conserving living things and the environment to sustain the balance in the ecosystem.

As population increases, humans put more pressure on the environment.

ICT opportunity

Type 'Human influences on the environment' on YouTube Kids. Then select and watch a video.

Write down all the ways that humans negatively affect the environment.

....

....

....

In this lesson, you will look at some of the ways that humans can damage the environment.

Deforestation

Make a note

Trees are cut down to provide materials and to provide room for urbanisation and agriculture.

Term 1 Unit 1 The environment

1. The table below shows some data about deforestation in Jamaica. Calculate the overall loss of forest area. .... km²

Area of forest lost (km²)	Area of forest grown (km²)
-393	+95

Source: https://rainforests.mongabay.com/deforestation/charts/latin-america/caribbean.html

2. One of the ways that areas are cleared of trees is a method called 'slash and burn'.

   a) Looking at the picture above, describe what is meant by 'slash and burn'.

   ....

   ....

   b) What effects do you think this will have on the animals living in the forest?

   ....

   ....

   c) What effects do you think this will have on air quality?

   ....

   ....

How humans damage the environment

Pollution

Other impacts caused by humans include pollution. Think of an area in your local community that has become polluted.

1 Describe the type of pollution. Is it water, air or land pollution?

....

2 What caused this pollution?

....

3 Suggest ways you think that this pollution could be reduced or stopped.

....

....

Quick quiz

Read each example and say whether it is land, water or air pollution.
1. Sewage washing into rivers.   ....
2. Rubbish tips.   ....
3. Burning forest.   ....
4. Oil tanker spills in the sea.   ....

Term 1 Unit 1 The environment

Conservation

> **Learning objectives**
> - Justify the importance of conserving the natural environment.
> - Be aware of my responsibility to preserve the environment.

Think about mangroves in Jamaica. They are special environments.

Write down all the animals and plants that live there. Choose one animal and draw a picture of it.

In this lesson, you will research a Jamaican environment and explain why it is important and how we can conserve it.

Conservation

Web task: Mangrove wetlands

ICT opportunity

Search on the internet to find out about the Jamaican mangrove environment. You could use www.nepa.gov.jm or www.noaa.gov as starting points for your research.

1 Use the information you found on the internet and your own knowledge to answer these questions.

a) What are mangroves?

....

....

b) Why are mangroves important?

....

....

....

....

c) What are the reasons that mangroves are being destroyed?

....

....

d) What protection is put in place to prevent mangroves from being destroyed?

....

....

Term 1 Unit 1 The environment

2 A developer has proposed building a new hotel next to the mangroves in Portland to bring more money into the parish.

Write a letter to the parish council to encourage them to reject this proposal. Include in your letter the importance of the mangroves.

Dear .... ,

....

....

....

....

....

....

....

....

....

....

Quick quiz

Are these statements **true** (T) or **false** (F)?

1 Laws can be put in place to conserve environments.	T	F
2 Coral reefs are not worth protecting.	T	F
3 Mangrove wetland areas can protect areas from flooding.	T	F
4 Urbanisation is one reason for habitat destruction.	T	F

Self-check

☺ I understand this well.

😐 I understand this but need more practice.

☹ I do not understand this yet.

Learning objectives	☺	😐	☹
I can formulate a definition of the environment.			
I can describe different factors in different environments as living and non-living.			
I can describe some ways to measure the different factors in environments.			
I can describe different environments.			
I can investigate features of different environments.			
I can evaluate information to place animals and plants in their natural environment.			
I can describe soil of my home environment.			
I can investigate soils of different environments.			
I can analyse particle size of soil.			
I can plan an investigation to test the water-holding capacity of different types of soil.			
I can write a method identifying the dependent, independent and control variables.			
I can identify some special features that enable animals to live in different environments.			
I can research the adaptations of one Jamaican animal.			
I can describe different animal adaptations in relation to their environment.			
I can identify some special features that enable plants to live in different environments.			

Term 1 Unit 1 The environment

I can research the adaptations of one Jamaican plant.			
I can describe different plant adaptations in relation to their environment.			
I can describe different environmental changes that occur and their impacts.			
I can recognise the need for and importance of conserving living things and the environment to sustain the balance in the ecosystem.			
I can describe some activities by humans that damage the environment.			
I can outline the effects of human activities on the environment.			
I can justify the importance of conserving the natural environment.			
I am aware of my responsibility to preserve the environment.			

Climate change

> **Learning objectives**
> - Formulate a simple working definition of climate change.
> - Give an example of climate change.
> - Use mathematical skills to analyse climate change.

Think about all the different types of weather you have experienced. Write them down.

....

....

Do you think that weather is the same thing as the climate?

....

....

In this lesson, you will work out a definition of climate change and be able to give some examples of climate change.

What is climate?

1 Read these children's definitions of climate.

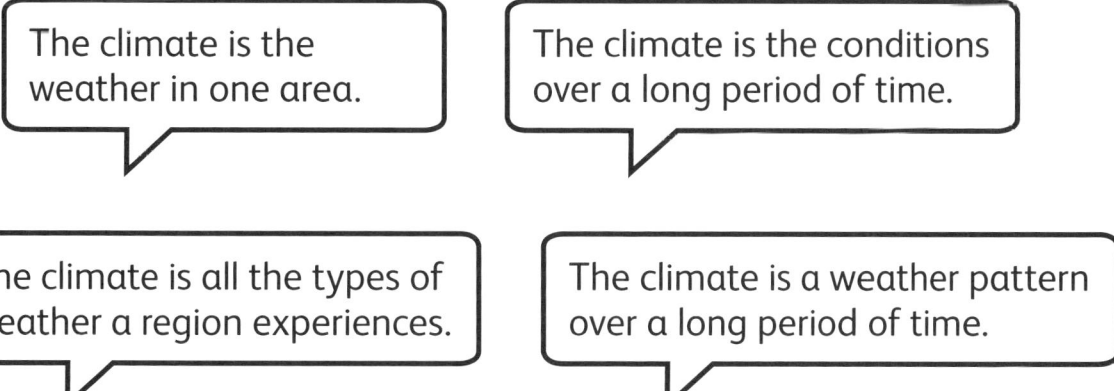

- The climate is the weather in one area.
- The climate is the conditions over a long period of time.
- The climate is all the types of weather a region experiences.
- The climate is a weather pattern over a long period of time.

2 Which do you think is the best definition? Can you improve on this definition?

Term 1 Unit 1 The environment

3 Write your definition in the space below.

....

....

Changing patterns in rainfall

One example of climate change is changing patterns in rainfall.

The table shows the mean rainfall in Jamaica for a year.

Month	Mean rainfall (mm)
January	108
February	85
March	83
April	138
May	226
June	184
July	135
August	181
September	205
October	271
November	185
December	148

Climate change

1 Plot the data on to the graph grid below as a bar graph and answer the questions.

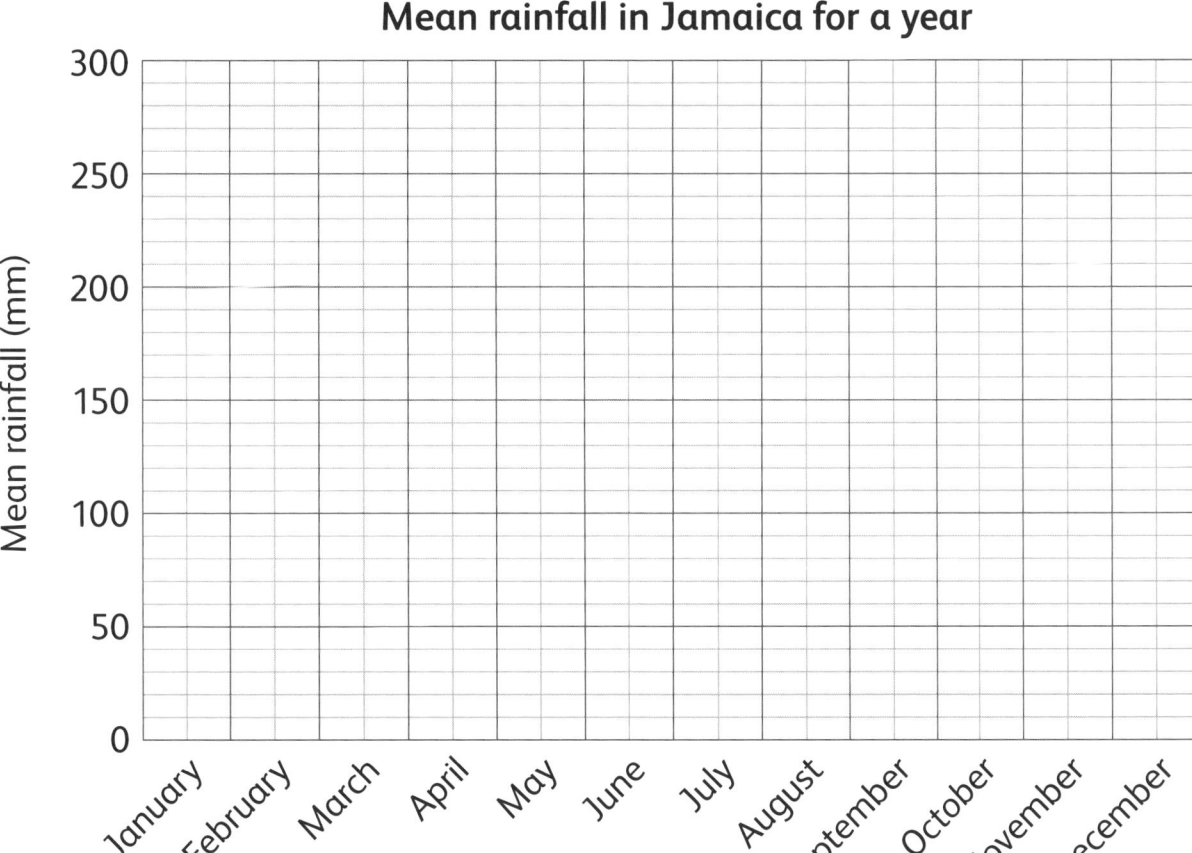

2 Answer these questions.

a) Which month has the highest rainfall? ....

b) Which month has the lowest rainfall? ....

c) Write down any trends you see in the results.

....

Quick quiz

Are these statements **true** (T) or **false** (F)?

1 Global warming is the cooling down of the Earth.   T   F
2 Global warming is an example of climate change.   T   F
3 Climate and weather are the same thing.   T   F
4 Global warming is caused by the greenhouse effect.   T   F

37

Term 1 Unit 1 The environment

Effects of climate change

> **Learning objectives**
> - Describe local, regional, international examples of climate change.
> - Explain the effects of climate change on humans.

1 Interview an elder member of your family. Ask if he or she thinks extreme weather like very heavy rainfall and hurricanes are more frequent now than they used to be.

Write down the responses in the space below.

....

....

....

In this lesson, you will use information about local examples of climate change and the effect they have on humans. You will use this information to create a poster.

2 Read the information sheet about climate change and its effects in Jamaica.

Changes to the climate in Jamaica have increased. There have been more severe hurricanes, an upturn in heavy rainfall, longer periods of drought and increased erosion of the shoreline.

Weather-related disasters have had an impact on Jamaica's economic growth. Hurricanes damage millions of dollars of crops. Powerful storms can damage roads and bridges, power and water supplies, as well as tourism businesses. Many coastal settlements are at risk from damage.

More intense rainfall events along with flooding can increase the spread of vector-borne and water-borne diseases. People can die or lose their homes.

Marine ecosystems are threatened by increasing temperatures and rising sea levels, affecting livelihoods and food sources.

Less rainfall and severe droughts reduce the availability of water resources, affecting agriculture and household use.

Effects of climate change

3 Use the information from the information sheet to create a poster on the effects of climate change in Jamaica. Things to include in your poster:
- a title
- examples of climate change that occur in Jamaica
- effects of climate change including economic effects, health effects, effects on society and infrastructure.

Quick quiz

1 Name two examples of climate change in Jamaica.
   a) ....   b) ....

2 Name two effects of climate change on humans.
   a) ....   b) ....

Term 1 Unit 1 The environment

The greenhouse effect

> **Learning objectives**
> - Describe the greenhouse effect.
> - Name some greenhouse gases.
> - Describe how humans enhance the greenhouse effect.

Have you ever been in a greenhouse? Research what the conditions are like in a greenhouse and what greenhouses are used for.

Write down what you have found out.

....

....

....

....

In this lesson, you will describe the greenhouse effect and describe how humans contribute to greenhouse gas emissions.

The greenhouse effect

Complete the sentences using these words to describe the greenhouse effect.

        atmosphere   Earth   Sun   radiation

a) Electromagnetic radiation from the .... passes through the Earth's atmosphere.

b) The .... absorbs most of the radiation and warms up.

c) The Earth radiates energy as infrared ....

d) Greenhouse gases in the .... absorb infrared radiation, which warms up the atmosphere.

The greenhouse effect

Greenhouse gases

> **Make a note**
>
> Humans are producing more greenhouse gases.

1 Name the two main greenhouse gases.

  a) …………………………… b) ……………………………

2 Name two human activities that release greenhouse gases into the atmosphere.

  a) ………………………………………………………………………………………

  b) ………………………………………………………………………………………

Carbon dioxide emissions

This graph shows carbon dioxide emissions.

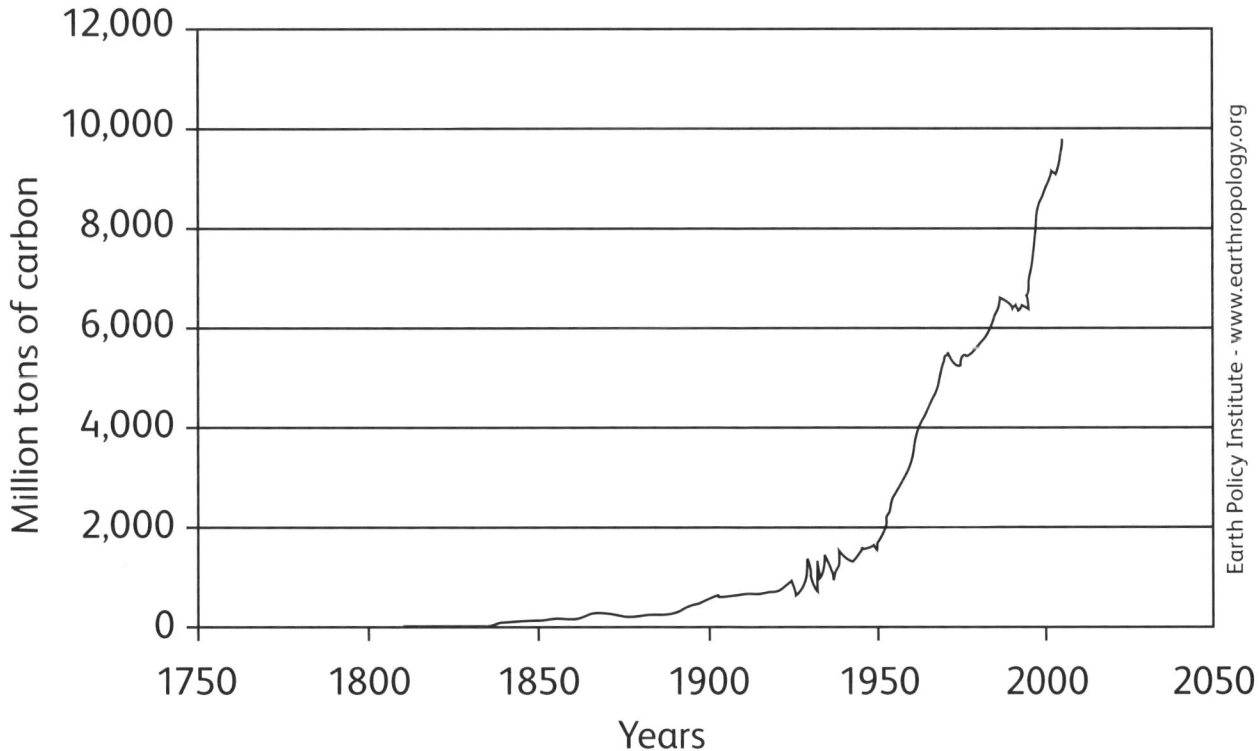

Source: EPI from BP; CDIAC; USGS

Term 1 Unit 1 The environment

1 When did we first start burning fossil fuels? ....

2 Which 50 years has seen the biggest increase in carbon dioxide emissions by fossil fuels? Tick (✓) the correct answer.

1800–1850 ☐   1850–1900 ☐   1900–1950 ☐   1950–2000 ☐

3 Suggest ways that we could reduce the amount of carbon dioxide emissions.

....

....

Quick quiz

Are these statements **true** (T) or **false** (F)?

1	Carbon dioxide is a greenhouse gas.	T F
2	Carbon dioxide emissions are decreasing.	T F
3	Burning fossil fuels releases greenhouse gases.	T F
4	Conserving energy reduces the emission of greenhouse gases.	T F

Causes and effects of soil degradation

> **Learning objectives**
> - Explain what 'soil degradation' means.
> - Describe the factors that cause soil degradation.
> - Describe the effects of soil degradation.
> - Describe ways to prevent soil degradation.

Find pictures of landslides online. Either draw or print out a picture and stick it in the space below.

Write down what happens when a landslide occurs.

....

....

....

Term 1 Unit 1 The environment

Write down what effects can landslides have on people.

....

....

....

In this lesson, you will find out what soil degradation is and the causes and effects of it.

ICT opportunity

Search on the internet to find out about soil degradation.

Use the information you found on the internet to answer the following questions.

1. What is soil?

    a) .... b) ....

    c) .... d) ....

2. What does soil degradation mean?

    ....

3. What are the three routes of soil degradation?

    a) ....

    b) ....

    c) ....

4. Give an example of each route of soil degradation.

    a) ....

    b) ....

    c) ....

Causes and effects of soil degradation

5 What is soil erosion?

....

....

6 Is Jamaica a region severely affected by soil degradation?

....

7 Describe the ways in which we can prevent soil degradation.

....

....

....

Soil erosion web exercise

Now we are going to focus on soil erosion.

ICT opportunity

Go online and search for 'Soil erosion and degradation' on the 'You matter' website.

Use the information you found on the internet and your own knowledge to complete the mind map.

Mind map

Term 1 Unit 1 The environment

Quick quiz

Are these statements **true** (T) or **false** (F)?

1. Soil can be degraded by chemical fertilisers. T F
2. Climate change has no effect of soil degradation. T F
3. Urbanisation can increase soil degradation. T F
4. Deforestation is one way we can prevent soil erosion. T F

Solid waste pollution

> **Learning objectives**
> - List sources of solid waste pollution.
> - Describe effects of solid waste pollution including spread of diseases.
> - Propose measures to reduce/eliminate selected sources of solid waste pollution.

Think of all the things that you throw in the bin in your home.

Write a list of items.

- ....
- ....
- ....
- ....
- ....
- ....

In this lesson, you will find out the different types of solid waste that humans produce.

1 Look at this table about what makes up solid waste in two countries and then answer the questions on the next page.

Type of solid waste	Percentage of the total solid waste	
	Jamaica	Trinidad and Tobago
food	55	46
paper	13	13
plastic	12	12
cardboard	4	7
glass	4	6
metal	5	7
other	7	9

Term 1 Unit 1 The environment

    a) Which country produces the largest percentage of food waste?

    ....

    b) What percentage of solid waste is made up of glass in Trinidad and Tobago? ....

    c) Which types of solid waste are the same percentage in both countries?

    ....

    d) Suggest what might be found in the 'other' types of solid waste.

    ....

2 Think back to the starter activity and the solid waste that you produce in your home. Think of ways that you can reduce the solid waste that you produce and write them down.

....

....

....

....

....

Quick quiz

1 Name three types of solid waste.

    a) ....   b) ....   c) ....

2 Suggest two ways we can reduce solid waste.

    a) ....   b) ....

3 Give one problem associated with improper disposal of solid waste.

....

Reducing waste

> **Learning objectives**
> - Propose measures to reduce/eliminate selected sources of solid waste pollution.
> - Be aware of my responsibility to carry out good environmental practices.

Look around your house for items that can be recycled. Check the packaging for the recycling logo.

Draw a labelled diagram of all the items you found in your house that can be recycled.

ICT opportunity

Go online and have a look at the information on the Jamaica Information Service (JIS). Search for 'Get the Facts – The Benefits of Recycling'.

Term 1 Unit 1 The environment

Use the information from the JIS website and your knowledge to answer these questions.

1. Name two problems that solid waste can cause.

   a) ....

   b) ....

2. Name three benefits of recycling.

   a) ....

   b) ....

   c) ....

3. Suggest two ways that parishes could encourage citizens to recycle their solid waste.

   a) ....

   b) ....

4. Design a leaflet in the space on the next page to encourage people to recycle more.

   Include in your leaflet the different problems with solid waste pollution and the benefits of recycling. You could even research recycling facilities in your area and include them in your leaflet.

Reducing waste

Quick quiz

Are these statements **true** (T) or **false** (F)?
1. Food waste is not a type of solid waste. T F
2. Food waste can be composted. T F
3. Recycling can reduce solid waste. T F
4. Improper disposal of solid waste can spread disease. T F

Term 1 Unit 1 The environment

Self-check

☺ I understand this well.

😐 I understand this but need more practice.

☹ I do not understand this yet.

Learning objectives	☺	😐	☹
I can formulate a simple working definition of climate change.			
I can give an example of climate change.			
I can use mathematical skills to analyse climate change.			
I can describe local, regional, international examples of climate change.			
I can explain the effects of climate change on humans.			
I can describe the greenhouse effect.			
I can name some greenhouse gases.			
I can describe how humans enhance the greenhouse effect.			
I can describe what soil 'soil degradation' is.			
I can describe the factors that cause soil degradation.			
I can describe the effects of soil degradation.			
I can describe ways to prevent soil degradation.			
I can list sources of solid waste pollution.			
I can describe effects of solid waste pollution including spread of diseases.			
I can propose measures to reduce/eliminate selected sources of solid waste pollution.			
I am aware of my responsibility to carry out good environmental practices.			

Extension activity

Aim: To explore how humans affect a local environment.

Task: Research a local environment and write a report on how humans affect this environment and what they can do to improve the environment.

Use the research questions below to help you.

Use paragraphs.

Write your report on the next page.

Questions:

1. Choose a local environment that humans have affected. It could be a beach, mangrove wetland, forest or school playing field.

2. Do a survey of the animals and plants that live in this environment. Do they have any special adaptations?

3. Write down the conditions in that environment, for example, temperature, rainfall, soil type.

4. Describe how humans have affected the environment. Is it polluted? Is there solid waste pollution? Have people built or planted crops on it?

5. Has this environment been affected by climate change? Can you describe how?

6. How can we conserve this environment? Give your suggestions.

7. Try to take photos of your environment and how humans have affected it. If you can't take photos, draw a picture.

Term 1 Unit 1 The environment

Photo or drawing

Report

....

....

....

....

....

....

....

....

Practice test

1. What is the best definition of the term 'environment'?
   a) all the living things in an area and the physical surroundings
   b) all the non-living things in an area
   c) the physical surroundings only
   d) the physical surroundings and all the living and non-living things

Look at the table. Use it to answer questions 2 and 3.

Environment	Rainfall	Amount of shade	Temperature
A	high	high	high
B	low	low	high
C	low	low	low
D	high	low	low

2. Which letter represents a tropical rainforest environment? Circle.

   A   B   C   D

3. Which letter represents a desert environment? Circle.

   A   B   C   D

A scientist investigated the range of water temperatures that different species of fish were able to survive in.

Her results are shown in the table on the next page. The temperature that the fish can survive in is shaded.

Term 1 Unit 1 The environment

Use the table to answer questions 4, 5 and 6.

Species	Temperature / °C															
	20	21	22	23	24	25	26	27	28	29	30	31	32	33	34	35
A																
B																
C																
D																
E																
F																

4 Which species is able to survive the greatest range of temperatures?

....

5 The fish farm wanted to stock all the species of fish.

Which temperature should the water be kept at?..................................

6 One of the scientists suggested that the fish farm stocks all the fish apart from species D.

Using data from the table, explain why this might be a good idea.

....

....

A student analyses soil particles. They use 100 g of soil and pass it through sieves with different size holes and measured the mass of soil that remained in the sieve. The following table shows their results.

Size of sieve / mm	Mass of soil retained / g
40.00	0.00
20.00	0.00
10.00	21.30
5.00	18.75
2.50	37.50
1.25	62.75
0.63	78.80
0.31	98.80

7 Which is the most likely size of the largest particle? Circle.

   a) 40.00   b) 15.00   c) 10.00   d) 0.20

8 Calculate the mass of soil that has a particle size of less than 0.31 mm.

....

9 Name the dependent variable in this investigation.

....

10 Which is an effect of increased rainfall? Circle.

   a) landslides   b) drought
   c) deforestation   d) earthquakes

11 What is the best description of the term 'slash and burn'? Circle.

   a) deforestation caused by cutting down trees
   b) pollution caused by burning rubbish
   c) pollution caused by release of sewage
   d) deforestation caused by cutting and burning trees

The table on next page shows different ways that humans can affect the environment.

Term 1 Unit 1 The environment

	Creating protected areas	Slashing and burning – Deforestation	Creating large-scale plant monocultures
Conserves environment	high	high	high
Increases soil erosion	low	low	high
Increases the release of greenhouse gases	low	low	low

12 Tick (✓) each of the human effects in the table that conserves the environment.

13 Underline each of the human effects in the table that increases soil erosion.

14 Circle each of the human effects in the table that increases the release of greenhouse gases.

15 Which of the following causes chemical soil degradation?
   a) deforestation
   b) urbanisation
   c) fertiliser pollution
   d) monoculture farms

16 Why it is better to reduce the production of waste rather than recycling waste?
   a) Recycling requires more energy to be used.
   b) Recycling is cheaper than reducing waste.
   c) New products can be made with recycling.
   d) Recycling waste reduces landfill.

Term 1 Unit 2 — Light and sound

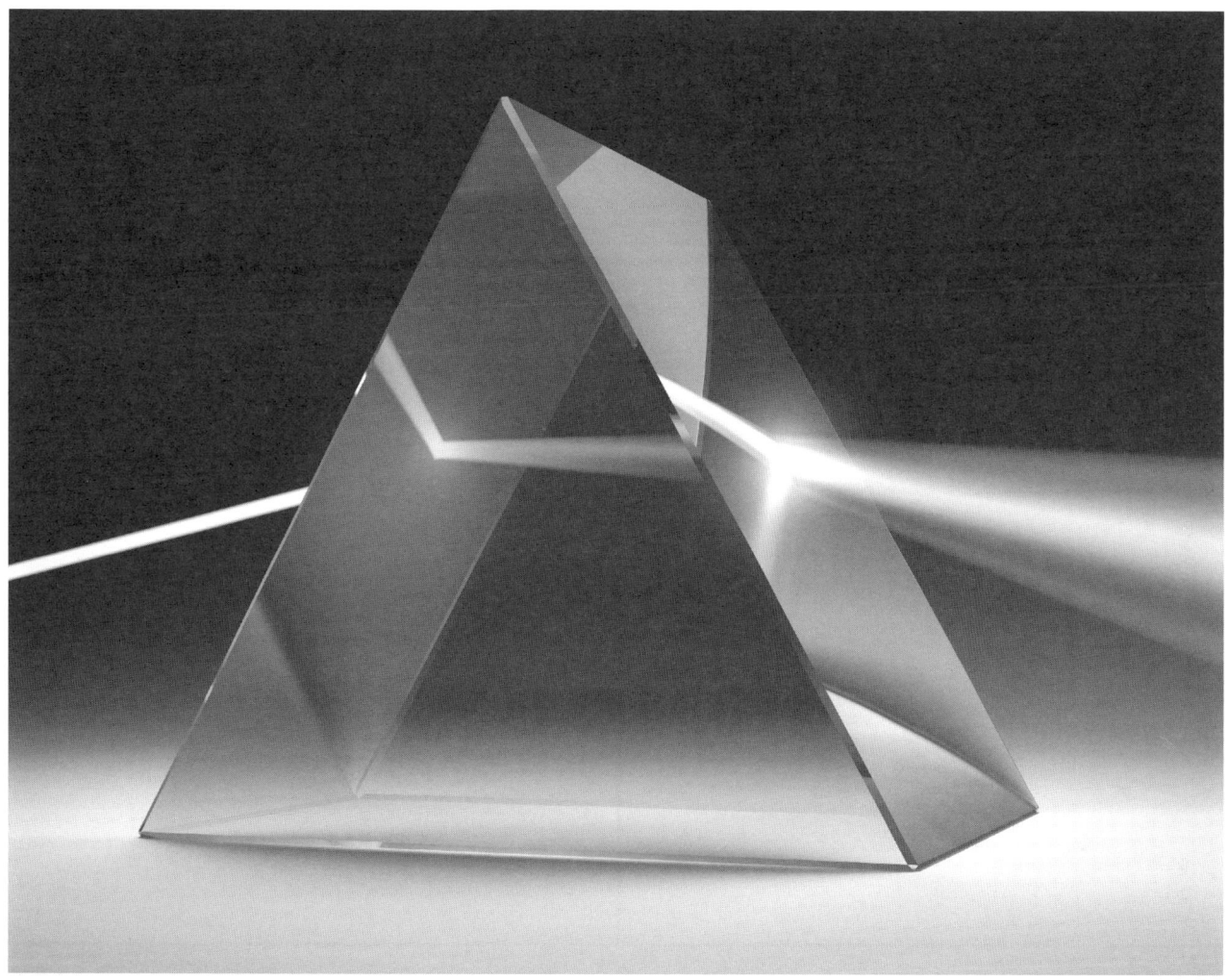

Luminous or non-luminous?

> **Learning objectives**
> - Describe what visible and non-visible objects are.
> - Describe what is meant by the terms 'luminous' and 'non-luminous'.
> - Distinguish between luminous and non-luminous objects/organisms.

Term 1 Unit 2 Light and sound

Think about all the things that you can see. Can you also think of things that you can't see?

Objects that you can see are called 'visible objects' and objects that you can't see are called 'non-visible objects'.

Complete the table to write down some examples of visible and non-visible objects. One example of each has been done for you.

Visible objects	Non-visible objects
chair	air

In this lesson, you will learn about what we mean by the terms 'luminous' and 'non-luminous' and be able to give examples of luminous and non-luminous objects.

Luminous and non-luminous

1. Think about some of the visible objects that you wrote down in the starter activity.

Luminous or non-luminous?

> **Make a note**
>
> We can see the visible objects because they either emit their own light or they reflect light. These are luminous and non-luminous objects respectively.

2. Write your own definition of the terms 'luminous' and 'non-luminous' objects.

   luminous objects:

   ....

   ....

   non-luminous objects:

   ....

   ....

Examples of luminous and non-luminous objects

1. Complete the wordsearch to find all the objects that follow below.

S	S	I	B	C	G	Y	W	E
T	B	T	O	R	C	H	A	P
A	D	F	O	S	T	R	W	E
R	P	A	K	Y	H	R	C	N
M	L	S	B	S	H	O	E	N
A	N	Q	U	R	V	C	H	L
E	M	O	O	N	E	K	D	C
L	I	G	H	T	B	U	L	B

book   lightbulb   moon   pen
rock   shoe   star   sun   torch

Term 1 Unit 2 Light and sound

2   Organisms can also produce light.

   a)   Name the word given to living things that produce light.

   ....

   b)   Write down one example of an organism that can be luminous.

   ....

> **Quick quiz**
>
> Are these statements **true** (T) or **false** (F)?
> 1   Some objects emit their own light.   T   F
> 2   It is possible for non-visible objects to emit their own light. T   F
> 3   You can see non-luminous objects in a dark room.   T   F
> 4   You can see luminous objects in a dark room.   T   F

How does light travel?

> **Learning objectives**
> - Describe how light travels.
> - Explain how shadows are formed.
> - Investigate the properties of light.

Stand outside on a sunny day. Notice where your shadow is and where the light of the sun is coming from.

Safety
Don't look directly at the sun.

Fill in the activity sheet below. (Note if it is not a sunny day this can be done inside using light from a lightbulb.)

Time of day: ....

Describe your shadow. Was it tall and thin or short? Was it behind you or in front of you when you were facing away from the sun? Write down your observations.

....

....

....

In this lesson, you will learn about how light travels and how shadows are formed.

Casting shadows

ICT opportunity

Watch a video showing how to make shadow pictures with your hands.

1. Get a lamp and shine it on to a light background like a wall.

2. Try to make as many shadow animals as you can. To do this you need to make the shapes using your hands in between the light source and the wall.

3. Write down the names of the shadow animals you managed to make.

Shadow puppet

1. Follow the instructions to make a shadow puppet and answer the following questions.

 **Safety**
Take care when using scissors.

a) Draw an outline of a simple shape on to a black card. Cut it out and stick it on to a stick to make a shadow puppet.

b) Set up a lamp shining on to a light background like a wall.

c) Put your puppet in between the lamp and the wall so that it creates a shadow.

d) Move your shadow puppet closer to the light and then away from the light.

e) Write down what you notice happening to the shadow as you move the puppet.

When I move the shadow puppet closer to the light, ...

How does light travel?

When I move the shadow puppet away from the light, ...

....

2   Circle the correct words in bold to explain how a shadow is formed.

a) Light travels in **curved** / **circular** / **straight** lines.

b) When we place an object in the path of the **air** / **light** / **sound**, it blocks it.

c) The area where the light is blocked behind the object will be **dark** / **light** and will form a **reflection** / **shadow**.

Solar eclipse

ICT opportunity

Go online and have a look at a solar eclipse happened in the past.

Use the information and answer the following questions.

1   When did the eclipse occur? ....

2   How is the eclipse formed? ....

....

....

3   Why shouldn't you look directly at the solar eclipse?

....

Quick quiz

Are these statements **true** (T) or **false** (F)?

1   Light travels in straight lines. T   F
2   Shadows are formed because light passes through objects. T   F
3   Solar eclipses are caused by the Moon passing in front of the Sun. T   F

Translucent, transparent, opaque

> **Learning objectives**
> - Describe what is meant by the terms 'translucent', 'transparent' and 'opaque'.
> - Give some examples of translucent, transparent and opaque objects.
> - Investigate the interaction of light with materials that are transparent, translucent and opaque.

The boxes on the left contain some key terms about light.

The boxes on the right show definitions of these terms.

Link each term to the correct definition.

Term
translucent
transparent
opaque

Definition
allows no light to pass through
allows some light to pass through
allows most of the light to pass through

In this lesson, you will learn about translucent, transparent and opaque objects.

Translucent, transparent and opaque objects

1. Collect a variety of objects from around your home. Make sure some of them are see-through.

2. Use a torch to shine light on to the object to see if the light passes through.

3. You will be able to classify each object as translucent, transparent or opaque depending on how much light passes through it.

Translucent, transparent, opaque

4 Fill in the table below with your results.

Name of object	How much light passes through	Translucent / transparent / opaque

Can you classify these objects?

Circle the correct word in bold to show whether each object is translucent, transparent or opaque.

1 air translucent / transparent / opaque

2 card translucent / transparent / opaque

3 cricket ball translucent / transparent / opaque

4 frosted glass translucent / transparent / opaque

5 glass translucent / transparent / opaque

6 wax paper translucent / transparent / opaque

7 water translucent / transparent / opaque

Quick quiz

Are these statements **true** (T) or **false** (F)?

1 All objects can be classified as translucent, transparent or opaque. T F

2 Translucent objects let most light pass through. T F

3 Glass is an example of an opaque object. T F

Term 1 Unit 2 Light and sound

Dull or shiny?

> **Learning objectives**
> - Describe what is meant by the terms 'dull' and 'shiny'.
> - Give some examples of dull and shiny objects.
> - Investigate the interaction of light with materials that are dull and shiny.

Look at your reflection in a mirror.

Describe the surface of the mirror in the space below.

....

....

In this lesson, you will learn about how light interacts with dull and shiny objects.

Dull and shiny objects

1   Collect a variety of objects from around your home.

2   Examine the surface of each object to see whether it is smooth or rough and whether it is reflective.

3   You will be able to classify each object as dull or shiny.

4   Fill in the table below with your results.

Name of object	Description of surface	Dull / shiny

Why does light reflect differently off dull and shiny surfaces?

1 Complete the sentences using these words to explain why light reflects differently off dull and shiny surfaces.

different   dull   reflected   regular   shiny

a) When light hits an opaque surface it can be ……………………………….

b) If the surface is smooth the light reflects in a ……………………………. way.

c) This gives the surface a ……………………………. appearance.

d) If the surface is rough the light reflects in ……………………………. directions.

e) This gives the surface a ……………………………. appearance.

2 Complete the two diagrams below to show how light reflects off a dull and shiny surface.

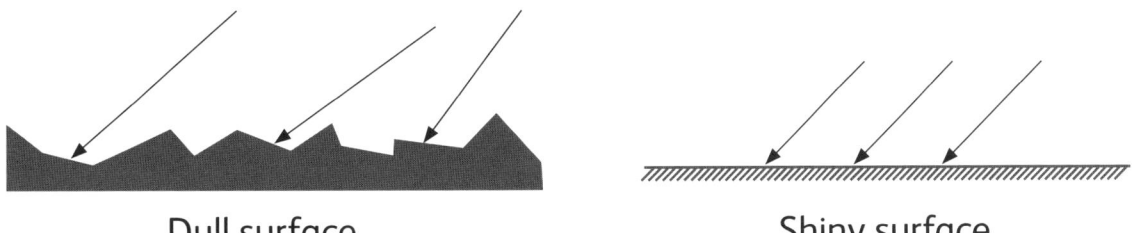

Dull surface   Shiny surface

Quick quiz

1 Name one example of a shiny object.   ………………………………

2 Name one example of a dull object.   ………………………………

3 Describe the difference in the way light reflects off a dull and shiny surface.

………………………………………………………………………………………………………

………………………………………………………………………………………………………

Term 1 Unit 2 Light and sound

Reflection 1

> **Learning objectives**
> - Investigate the pathway of light reflected off a mirror.
> - Investigate the interaction of light with mirrors.

Look around your house.

Write down the number of reflective objects you find in each room.

Room	Number of reflective objects	Names of reflective objects

In this lesson, you will learn about how light interacts with a reflective surface.

Investigating the pathway of light

1. Collect a torch, a piece of black card, a mirror and some modelling clay.

2. Cut a thin slit in your black card and stand it up using the modelling clay.

3. Shine the torch through the black card and on to the mirror. The ray of light should reflect off the mirror.

4. Move the position of the torch. You will see the angle of the incident ray and the angle of the reflected ray move.

Safety
Take care when using scissors.

Make a note

The ray of light from the torch to the mirror is called the 'incident ray' and the ray of light reflected from the mirror is the 'reflected ray'.

Reflection 2

> **Learning objectives**
> - Investigate some effects of reflection in everyday life.
> - Use the laws of reflection to make a model periscope.

Think of some examples where we use mirrors in everyday life.
Write some of these examples down in the space below.

....

....

....

In this lesson, you will investigate how we can use reflection.

Making a model periscope

These are instructions on how to make a periscope. You might need an adult to help you with the cutting.

You will need:
- 2 juice or milk cartons
- 2 small rectangular mirrors
- a craft knife
- masking tape
- a ruler.

! Safety
Take care when using a craft knife.

Term 1 Unit 2 Light and sound

1 Follow the instructions to make your periscope.

Instructions	Diagram	
1 Cut off the peaked top of the carton.	1	2
2 Cut a square on one side of the carton at the bottom. Leave approximately 2–3 cm from the edge of the carton.		
3 Turn the carton on its side and draw a line at a 45° angle from the bottom right-hand corner.		
4 Cut this line and slide the mirror into the carton so the reflective side is facing the hole you cut. You can use masking tape to keep the mirror in position.	3	4
5 When you look through the hole, you should be able to see the ceiling. If you can't, try adjusting your mirror slightly until you can.		
6 Repeat steps 1–5 with the other milk carton.	7	8
7 Stand up the two cartons, one with the hole facing you and one with the hole facing away from you.		
8 Place the carton with the hole facing away from you on top of the other carton and tape them together.		
9 You should now be able to use your periscope to look over things taller than you and to look around corners.		

2 Write an explanation about how your periscope works.

Try to include the words: **reflection**, **angle**, **incidence**, **mirror**.

....

....

....

....

....

Refraction

Quick quiz

1. How many mirrors does a periscope use? ....
2. What angles do the mirrors need to be in a periscope? ....
3. If the light hits the mirror in a periscope at a 50° angle, what angle will the light be reflected off the mirror? ....

Refraction

Learning objectives
- Describe what is meant by the term 'refraction'.
- Investigate some effects of refraction in everyday life.

Draw an arrow on a plain piece of paper.

Stand your piece of paper behind a clean glass.

Fill the glass up with water.

What do you observe?

Draw a picture of the arrow before and after you add the water in the space below.

Term 1 Unit 2 Light and sound

Refraction activity

1 Place a straight pencil into a glass half full of water.

2 Look at the pencil from the top. Does the pencil still look straight?

3 Now look at the pencil from the side. Does it still look straight?

4 Draw a picture in the box provided.

5 Circle the correct words in bold in the sentences below to describe what is happening in your experiment.

   a) When light passes from one medium to another, it changes **direction** / **shape**.

   b) This is called **reflection** / **refraction**.

   c) The light takes longer to pass through a **liquid** / **gas** than it does a **liquid** / **gas**.

   d) The light passing through the water is refracted further, giving the pencil a **bent** / **straight** appearance.

   e) This is because liquids are **denser** / **less dense** than gases.

Quick quiz

Are these statements **true (T)** or **false (F)**?

1 Light travels at different speeds through different media.   T   F

2 Light travels faster in liquids than in gases.   T   F

3 Light is refracted further when travelling through a denser medium.   T   F

Lenses

> **Learning objectives**
> - Describe different types of lenses and what they are used for.
> - Describe the use of different types of lenses in everyday life.

Try and think of all the everyday objects that use lenses.
Write a list in the space provided.

....

....

In this lesson, you will learn about how lenses work and some examples of their use in everyday life.

Lenses web activity

ICT opportunity

Go online and find information about the lenses.

Use the information and your knowledge to answer the following questions.

1 What is a lens?

   ....

2 What are lenses used for?

   ....

Term 1 Unit 2 Light and sound

3 Draw a picture of a convex lens in the box.

4 How does light travel through a convex lens?

....

....

5 What are convex lenses used for?

....

6 Draw a picture of a concave lens in the box.

7 How does light travel through a concave lens?

....

....

....

8 What are concave lenses used for?

....

....

9 Describe the vision of near-sighted people. ....

....

10 Describe the vision of far-sighted people. ....

....

11 What type of lens is used to correct near-sightedness? ....

12 What type of lens is used to correct far-sightedness? ....

13 Glasses are used by people so that the light is focused on to which part of the eye?

....

Investigating glasses

Do you or anyone you know use glasses?

See if you can examine other people's glasses. Ask them whether they are near- or far-sighted and see if you can identify the type of lenses that are used in their glasses.

Write down your observations.

....

....

....

Quick quiz

Are these statements **true (T)** or **false (F)**?

1 Microscopes use lenses.   T   F
2 Convex lenses make objects appear larger than they are.   T   F
3 Near-sighted people use glasses with concave lenses.   T   F
4 The retina is the name given to the lens in the eye.   T   F

Term 1 Unit 2 Light and sound

Self-check

Learning objectives	😊	😐	☹
I can describe what visible and non-visible objects are.			
I can describe what is meant by the terms 'luminous' and 'non-luminous'.			
I can distinguish between luminous and non-luminous objects/organisms.			
I can describe how light travels.			
I can explain how shadows are formed.			
I can investigate the properties of light.			
I can describe what is meant by the terms 'translucent', 'transparent' and 'opaque'.			
I can give some examples of translucent, transparent and opaque objects.			
I can investigate the interaction of light with materials that are transparent, translucent and opaque.			
I can describe what is meant by the terms 'dull' and 'shiny'.			
I can give some examples of dull and shiny objects.			
I can investigate the interaction of light with materials that are dull and shiny.			
I can investigate the pathway of light reflected off a mirror.			
I can investigate the interaction of light with mirrors.			
I can investigate some effects of reflection in everyday life.			
I can use the laws of reflection to make a model periscope.			
I can describe what is meant by the term 'refraction'.			
I can investigate some effects of refraction in everyday life.			
I can describe different types of lenses and what they are used for.			
I can describe the use of different types of lenses in everyday life.			

What is sound?

> **Learning objectives**
> - Describe how sound is made.
> - Investigate the properties of sound.

Shut your eyes and listen to all the sounds that you can hear in two minutes.

Can you also identify the sources of these sounds (where they are coming from)?

Write down what you hear.

....

....

....

In this lesson, you will learn about how sound is made.

Sound vibrations

1. Listen to some music or the radio through a speaker.
2. Gently put your fingertips on the speaker.
3. Write down what you can feel.

....

Make a string telephone

You will need:
- two plastic or paper cups
- string
- a craft knife.

! Safety

You will need an adult for this activity to help you make holes with a knife.

Term 1 Unit 2 Light and sound

Method

1. Get two plastic or paper cups, some string and a craft knife.

2. Place a small hole in the bottom of each cup.
3. Thread the string through the cup and secure it with a knot.
4. You should have a string telephone that looks like this.
5. With a friend, stand a distance apart so the string is tight and take turns in talking into the cup and listening.
6. You can experiment with longer and shorter pieces of string.
7. Gently touch the string while one of you is talking. What do you notice? Write down your ideas.

....

....

Can sound travel through a vacuum?

ICT opportunity

Watch a video about the Bell Jar Experiment to find out how sound travels.

What is sound?

The air particles are being removed from the jar the bell is in.

1 What does this tell you about how sound travels?

....

2 Use the results from the experiments you have done and your own knowledge to answer the following questions.

   a) What organ of our body do we use to hear sound? ....

   b) How is sound made? ....

   c) If someone is talking to us, how does the sound reach us so we can hear it?

   ....

   ....

   d) Can sound waves travel through a vacuum? Why or why not?

   ....

   ....

3 Write down a sound that you like hearing and one that you don't. Can you explain why you like or don't like these sounds?

....

....

Quick quiz

Are these statements **true** (T) or **false** (F)?
1 Sound is caused by vibrations.   T   F
2 We sense sound with our fingertips.   T   F
3 Sound needs particles to travel.   T   F

Term 1 Unit 2 Light and sound

Pitch

Learning objectives
- Describe what causes sounds of different pitches.
- Investigate the properties of sound.

ICT opportunity

Watch and listen to a video that demonstrates sound frequency.

Safety

Do not listen to this through earphones. If you do, have the volume set low.

What is frequency measured in? ....

How does the sound change with increasing frequency? ....

What is the lowest frequency sound that you can hear? ....

What is the highest frequency sound that you can hear? ....

What is the normal audible range for humans? ....

In this lesson, you will learn about how sound of different pitches is made.

Cardboard guitar

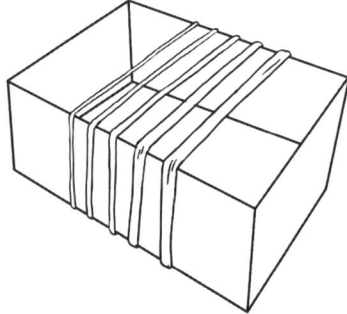

You will need:
- a cardboard cereal box (or similar)
- a range of rubber bands
- a craft knife
- scissors.

Safety

You will need an adult for this activity to help you with the cutting.

82

Pitch

Method

1. Cut a square hole in one side of the cereal box.
2. Arrange your elastic bands from thickest to thinnest on the box over the hole so it looks like this.
3. Pluck the bands one at a time and listen to the sounds they make.
4. Write down your observations as the bands get thinner.

   ....

   ....

5. Place a pencil over the bands to make the bands that you pluck shorter. Pluck each band again and listen to the sounds that each one makes.
6. Write down your observations as the bands got shorter.

   ....

   ....

How does sound change in pitch?

Fill in the gaps in the sentences to describe how sound changes in pitch.

1. Sounds that have a .... pitch have a high ....

2. This means that sound vibrates at a .... rate.

3. Sounds that have a .... pitch have a low ....

4. This means the sound vibrates at a .... rate.

5. The longer or wider a string is on a musical instrument the .... the pitch.

Term 1 Unit 2 Light and sound

Musical instruments and pitch

Look at the picture of musical instruments.

Double bass **Cello** **Violin**

1 Put them in order of pitch from lowest to highest.

....

2 Give a reason why you put them in this order.

....

....

....

Quick quiz

Are these statements **true** (T) or **false** (F)?
1 Pitch is the volume of the sound. T F
2 A higher pitch is caused by a higher frequency. T F
3 Longer stringed instruments results in a higher pitch. T F

Speed of sound

> **Learning objectives**
> - Explain how the speed of sound changes in different media.
> - Analyse data to describe how the speed of sound changes in different media.

ICT opportunity

Go online and listen to the whale song.

Whales produce low frequency sound. It is thought it can travel up to 10 000 miles.

Do you think sound in air can travel 10 000 miles? ....

In this lesson, you will learn about how the speed of sound changes in different media.

Speed of sound data analysis task

Look at this table of how quickly sound travels through different materials and answer the questions on the next page.

Material	Solid / liquid / gas	Speed of sound
oxygen	gas	332
nitrogen	gas	354
oil	liquid	1 460
water	liquid	1 510
granite	solid	5 400
steel	solid	5 980

Term 1 Unit 2 Light and sound

1. Plot the data as a bar graph on the graph paper below. Colour the blocks for the gases, liquids and solids in different colours.

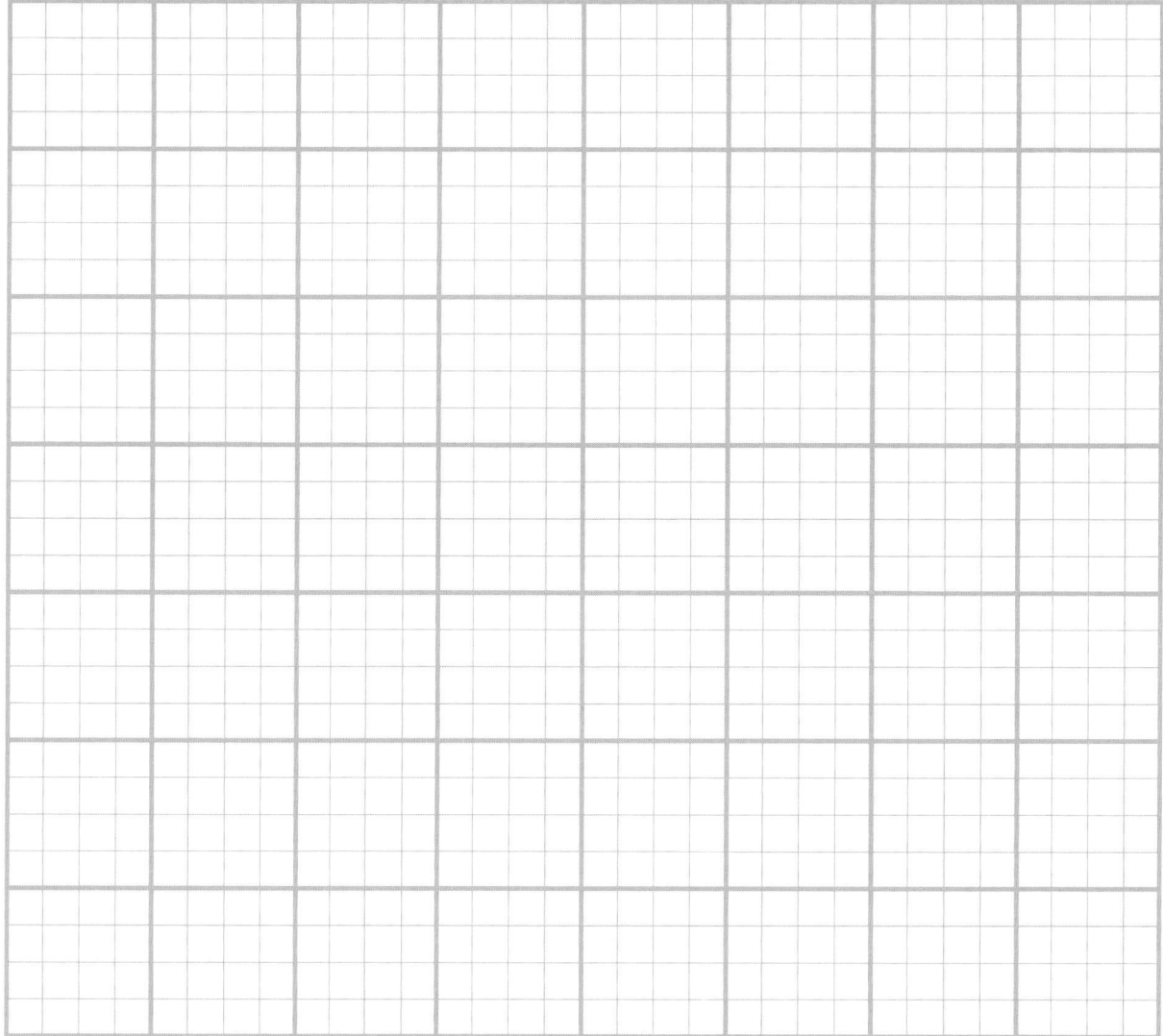

2. In which material did sound travel the fastest? ....

3. In which material did sound travel the slowest? ....

4. What are your observations about if a material is a solid, liquid or a gas and the speed of sound?

   ....

   ....

5 Can you explain a reason for your answer to question 4? Think about the density of particles.

....

....

> **Quick quiz**
>
> Circle the correct answers.
> 1. Sounds travel fastest in **solids / liquids / gases**.
> 2. The speed of sound is measured in **metres per second / miles per hour / kilometres per hour**.
> 3. Sounds travel slower in **less / more** dense materials.

Volume

> **Learning objectives**
> - Describe how sound is measured.
> - Analyse data to investigate sounds of different volumes.

Shut your eyes and listen carefully.

What is the loudest sound you can hear? ....

What is the quietest sound you can hear? ....

In this lesson, you will learn about how volume is measured the volume of different sounds.

Term 1 Unit 2 Light and sound

Volume of different sounds

Look at this table of the volume of different sounds and answer the following questions.

Sound	Volume / decibels
silence	0
whisper	20
normal conversation	50
hairdryer	70
lawnmower	90
chainsaw	110
stadium crowd noise	130
airplane taking off	140
explosion	170

1 What unit is volume measured in? ....

2 If a sound was 80 decibels, would it be louder or quieter than a hairdryer?

   ....

3 Suggest what volume a quiet conversation would be. ....

4 How many times louder is an airplane taking off than a hairdryer?

   ....

5 An American alligator can make a sound of 90 decibels. It is as loud as which sound from the table? ....

.... Volume

6 A howler monkey can make a sound of 140 decibels. It is as loud as which sound from the table?

....

Loud or soft sound?

Look at the list of different sounds and classify them into loud and soft in the table below.

someone breathing　　door slamming　　inside a library
mosquito buzzing　　vacuum cleaner　　alarm clock
ambulance siren　　rustling leaves

Loud sounds	Soft sounds

Quick quiz

Are these statements **true (T)** or **false (F)**?

1. Sound measures the pitch of a noise.　　T　F
2. A loud sound has a lower volume.　　T　F
3. The sound of someone breathing will have a low number of decibels.　　T　F

Term 1 Unit 2 Light and sound

Noise pollution

> **Learning objectives**
> - Explain why sounds may be interpreted as pleasant/unpleasant.
> - Identify sources of noise pollution, and ways to eliminate them.

Look at the list of sounds. Do you find them pleasant or unpleasant? Circle.

Lawn mower	**unpleasant / pleasant**
Music	**unpleasant / pleasant**
A baby crying	**unpleasant / pleasant**
A stream running	**unpleasant / pleasant**

Explain why you find some of the sounds pleasant and some of these sounds unpleasant.

....

....

In this lesson, you will learn about why some sounds are pleasant and some sounds are unpleasant. You will also be learning about sources of noise pollution.

Analysing building materials to reduce noise pollution

The members of a family live next to Kingston airport and are tired of the noise from the planes taking off and landing. They want to insulate their home to reduce the noise levels.

This table shows a list of the materials they could use.

Noise pollution

Material	Cost / Jamaican dollars	How much sound will be eliminated / decibels	Length of time before it needs replacing / years
A	17 000	10	30
B	16 000	10	60
C	15 000	25	15
D	22 000	40	30
E	20 000	30	20

1 Which material do you think they should insulate their home with?

  ....

2 Give a reason for your answer to question 1.

  ....

  ....

  ....

3 Suggest what other factors the family should consider before choosing an insulating material for their home.

  ....

  ....

Noise pollution or not?

1 Read each scenario on the next page.

2 For each scenario, say whether you think this should be classified as noise pollution and suggest a solution to reduce the noise.

Term 1 Unit 2 Light and sound

Scenario 1

A person plays in a steel drum band. They practise their music late at night for two hours every day. This annoys their neighbour who has children that go to school early each morning.

....

....

....

Scenario 2

A hotel has started to have live music entertainment every night for holiday-makers. Local residents are complaining about the noise of the entertainment because they are finding it hard to sleep.

....

....

....

Quick quiz

1. What is noise pollution?

   ....

2. Describe some ways that people can reduce noise pollution in their environment.

   ....

   ....

   ....

   ....

   ....

Self-check

Learning objectives	☺	😐	☹
I can describe how sound is made.			
I can investigate the properties of sound.			
I can describe what causes sounds of different pitches.			
I can explain how the speed of sound changes in different media.			
I can analyse data to describe how the speed of sound changes in different media.			
I can describe how sound is measured.			
I can analyse data to investigate sounds of different volumes.			
I can explain why sounds may be interpreted as pleasant/unpleasant.			
I can identify sources of noise pollution and ways to eliminate them.			

Term 1 Unit 2 Light and sound

Extension activity

Aim: To explore light and sound during a firework show

Task: Describe the light and sound science involved in a firework show and use this to produce a poster.

ICT opportunity

Use the research questions below to help you. Research using the internet and use ideas that you have learned during this lesson. You can make notes on the next page.

Draw your poster on page 96.

Questions:

1. How does light travel from the firework to us?
2. How do we see the firework?
3. Is it luminous or non-luminous?
4. Can you find out how fast the light travels?
5. How loud is a firework? Is it harmful? Does it count as noise pollution?
6. How does the sound travel? How do we hear the firework?
7. Can you find out how fast the sound of the firework travels?
8. Does the light and the sound from the firework reach us at the same time?

Extension activity

Notes

Term 1 Unit 2 Light and sound

Poster

Practice test

1 Which of these is a luminous object? Circle your answer.

   a) Earth   b) Moon   c) Sun   d) Venus

2 In the diagram, a person is standing and his shadow is formed from one side.

Where is the light of the sun coming from?

This table shows the results of the amount of light that different objects allow through.

Object	Amount of light transmitted through the object
A	most
B	all
C	none
D	most

3 Which objects can be described as translucent? ....

4 Which object is most likely to be made of glass? ....

Term 1 Unit 2 Light and sound

Look at this diagram of light reflecting off an object.

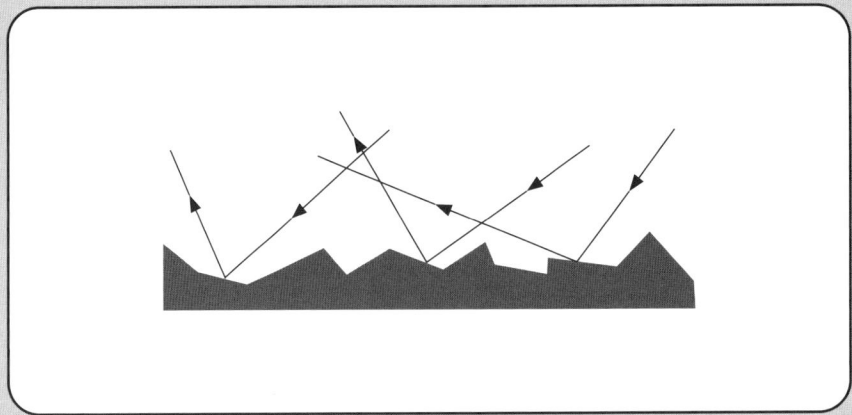

5 Which words can be used to describe the object? You can circle more than one.

a) shiny b) dull c) transparent
d) translucent e) opaque

Look at the diagrams of light travelling through a glass block.

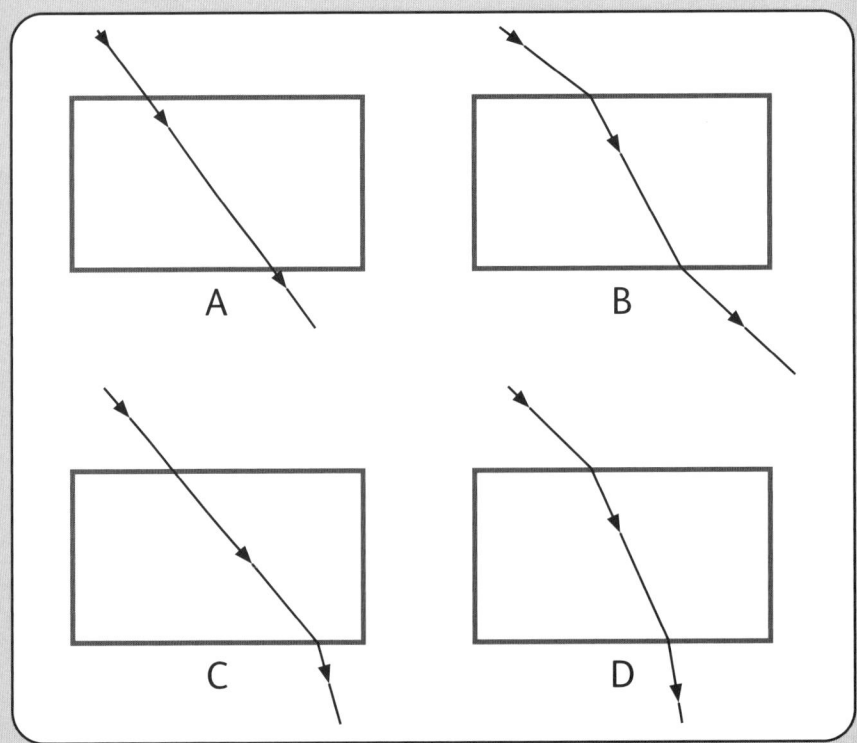

6 Which diagram shows the correct pathway of light? ....

7 Which statements explain why this occurs?

a) Glass is a reflective surface.

b) Light travels at different speeds through different media.

c) Light travels more slowly through gases.

d) Light has been refracted.

Use this table that shows some features of concave and convex lenses to answer questions 8 and 9.

Lens	Refracts light	Used to correct near-sightedness	Reflects light	Causes light rays to bend towards each other
concave				
convex				

8 Tick (✓) all the features in the table that concave lenses have.

9 Tick (✓) all the features in the table that convex lenses have.

A student investigated different noises. This table shows a summary of his results. Use this table to help you answer questions 10, 11 and 12.

Sound	Frequency / Hz	Volume / decibels
A	100	50
B	3 500	90
C	30 000	40
D	1 500	150
E	15 000	20

10 Which sound has the lowest pitch? ....

11 Which sounds could be harmful? ....

Term 1 Unit 2 Light and sound

12 Explain why humans can't hear sound C.

....

This table shows the speed of sound as it travels through different materials.

Object	Speed of sound m/s	Solid / liquid / gas
A	1 500	liquid
B	330	
C	6 500	
D	4 500	solid

13 Complete the table to show whether objects B and C are gases, liquids or solids.

Different types of materials

Term 2 Unit 1
Materials and their properties

Term 2 Unit 1 Materials and their properties

Different types of materials

> **Learning objectives**
> - Identify different materials.
> - Describe proper storage of materials.

Think about all the things made out of different materials in your house.

Go for a learning walk around your house and write down as many different types of materials as you can find in the space below.

In this lesson, you will learn about different types of materials and how we can classify them.

Different types of materials

Different materials

Complete the wordsearch to find all the materials below.

L	M	P	A	D	G	G	H	C
O	C	L	A	Y	L	S	R	A
W	O	A	C	P	A	P	E	R
Y	N	S	A	M	S	R	W	D
S	C	T	H	E	S	D	J	B
P	R	I	N	M	Z	B	K	O
E	E	C	S	A	W	K	L	A
R	T	M	L	I	O	C	E	R
S	E	G	F	W	O	O	L	D
M	E	T	A	L	D	U	R	S

wood glass clay
concrete paper wool
plastic cardboard metal

Storage of different materials

Make a note

Different materials require different methods of storage to keep people safe.

103

Term 2 Unit 1 Materials and their properties

Read these labels from products found at home and answer the following questions.

Product A

Product B

1 Which product is most likely to be a food stuff? ....

2 Suggest where product A should be stored. ....
   ....

3 Explain why people shouldn't smoke around product A.
   ....

Quick quiz

Match the symbols to these meanings.

damages the environment   explosive   flammable   toxic

A .... B .... C .... D ....

Properties of different materials

> **Learning objectives**
> - Identify properties of different materials.
> - Classify materials into groups using their properties.

Make a note

Materials have different properties that are useful for making different products.

Name the materials that these products are usually made out of.

electrical wire ....

cooking pot ....

baby toy ....

drinking cup ....

Identifying properties

Look at these materials.

Write a description of each material and its properties.

Try to use words like: hard / soft, stiff / flexible, light / heavy, opaque / transparent / translucent, if it conducts heat or electricity.

Material	Properties
metal	
glass	
wool	
plastic	
wood	

Term 2 Unit 1 Materials and their properties

Properties of objects at home

Choose two objects from your home environment.

Draw a picture or take a photo of your object. Write down the name of the object, the job the object does and the material it is made out of. Do you think it could be made out of any alternative materials?

Object 1:

....

Function:

....

Material:

....

Alternative materials:

....

Object 2:

....

Function:

....

Material:

....

Alternative materials:

....

Quick quiz

Circle the correct property in bold for each material.
1. Cotton — **can conduct electricity / is magnetic / is flexible.**
2. Iron metal — **can be magnetised / is soft / is transparent.**
3. Glass — **is soft / is flexible / is transparent.**

Investigating properties

> **Learning objectives**
> - Analyse properties of materials.
> - Link products with their properties.

Choose one object from your home environment.

Take a photo or draw a picture of the object in the box and write down its function and its properties.

Object:

....

Function:

....

Properties:

....

....

....

Analysing properties

A businessman has approached you to make some towels to sell in the shops by the beach. He wants to market the towels as being able to absorb the most water when compared to other towels.

He has narrowed down the different materials he would like to use and tested them.

A square of each material was put in a funnel and 100 cm³ of water poured on it.

A beaker collected the water that dripped from the material.

You will analyse the results to see which is the best material.

Term 2 Unit 1 Materials and their properties

Results

Material	Volume of water collected/cm³
A	75
B	80
C	20
D	45

1 Which material is the best for making a towel? ....

2 How much water did the best material absorb? ....

3 Why is it important to make sure that 100 cm³ of water was used each time?

   ....

4 Apart from the volume of water used, what other variables would need to be controlled in this investigation?

   ....

5 Suggest what other properties would need to be considered when choosing material for a towel.

   ....

Safety
What safety considerations need to be taken into account during his investigation?

Relating properties to uses

Match the property to the product

The boxes on the left show some products. The boxes on the right show some properties.

Draw lines to link the product with the property.

Product
frying pan
umbrella
hammer
rubber band

Property
waterproof
hard
flexible
conducts heat

Quick quiz

Name one property for each of these products.
1 raincoat   ....
2 kitchen knife   ....
3 tyre   ....

Relating properties to uses

Learning objectives
- Identify a range of objects and their uses.
- Identify the best object for a use dependent on their properties.

Term 2 Unit 1 Materials and their properties

Look at the objects in the box.

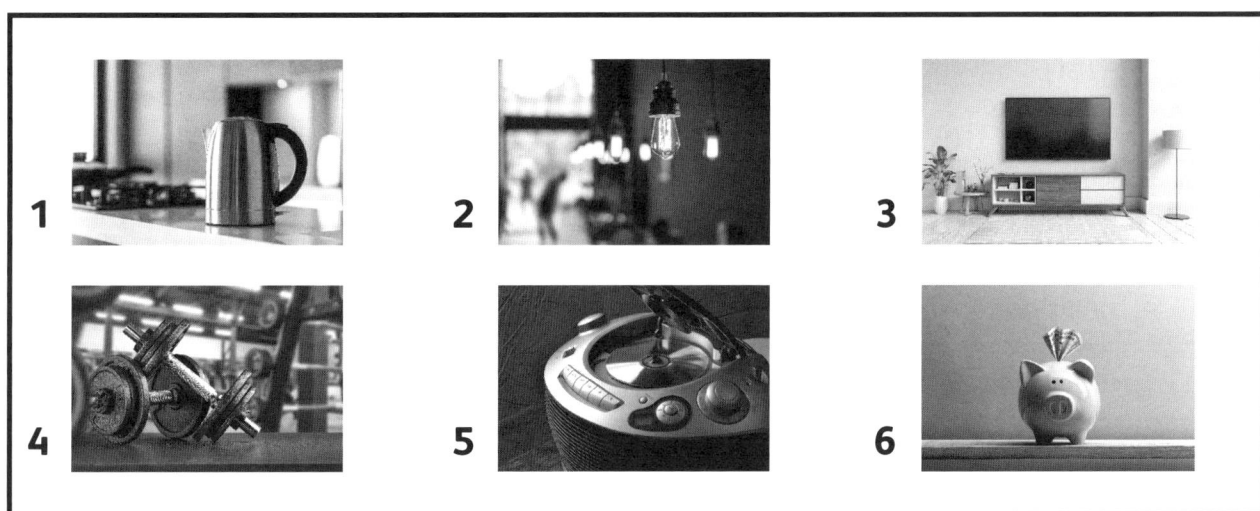

Write down the functions of the six objects.

1 ....

2 ....

3 ....

4 ....

5 ....

6 ....

Matching properties to their uses

Make a note

Different objects will have different properties that make them more useful for their function.

Relating properties to uses

Different objects will have different properties that make them more useful for their function.

Go around your home and pick up five everyday items.

Complete the table to show each object's properties and function. Can you add a sentence for each object to explain why its properties make it good for its function?

The first one has been done for you.

Object	Main property	Function	How is its property linked to its function?
1 frying pan	metal bit conducts heat	cook food	the metal conducts heat from the stove or hotplate to the food, heating the food up
2			
3			
4			
5			
6			

Term 2 Unit 1 Materials and their properties

Design a bag

You need to design a new school bag. Think about its function, what properties it would need and what materials you would use.

Draw your school bag in the space below.

Include labels showing the different features of your bag, what properties it has, what material it is made out of and how this relates to its function.

Quick quiz

Choose the most suitable material for each of these objects:

glass   metal   concrete   rubber   cotton   paper

1 drinking cup   ....

2 shopping bag   ....

3 pair of scissors   ....

Disposing of materials

Learning objectives
- Describe the ways that materials are disposed of.
- Explain the impacts of different methods of disposal.

Make a note

Some materials are recyclable.

Choose the kitchen or the bathroom in your house.

Look at all the products that end up being disposed of. Think about what they are made from and where they end up.

In the table below, write a list and say whether the products that you have found can be recycled, put in food waste or put in landfill.

Product	Material it is made from	Recycled / food waste / landfill

Life cycle of a plastic bottle

ICT opportunity

Watch a video about the lifecycle of a plastic bottle and the ways that it can be disposed of.

Term 2 Unit 1 Materials and their properties

Use the information in the video to write a piece of persuasive writing about why we should recycle plastic.

Things to include are:
- How plastics are made.
- Three ways that plastics can be disposed of.
- The environmental impacts of the three different methods of disposal.

Recycling

Quick quiz

1 Describe one way that plastics can affect the environment if they get into the sea.

....

....

....

2 Describe one environmental problem of disposing rubbish in landfill.

....

....

....

3 How long does it take for plastics to decompose?

....

....

....

Recycling

Learning objectives
- Identify materials that can be recycled at home.
- Describe the benefits of recycling.

Make a note

Single-use plastics can't be recycled. These include straws, coffee cups from takeaway cafés and plastic bags.

Term 2 Unit 1 Materials and their properties

Move through each room in your house and make a tally of how many single-use plastics there are in each room.

Room	Number of single-use plastics

Recycling web activity

ICT opportunity

Go online and find information about solid waste and recycling in Jamaica.

Use the information to answer the following questions.

1 How many tonnes of solid waste does Jamaica produce in a day?

   ....

2 How many plastic bottles were improperly disposed of?

   ....

3 Describe three reasons why we should recycle.

   a) ....

   b) ....

   c) ....

4 Describe three ways we can reduce the amount of solid waste produced.

   a) ....

   b) ....

   c) ....

Recycling

Recycling facilities

ICT opportunity

Go online and find information about your local recycling facility.

1. Write down the name and address of your nearest recycling facility and what it recycles.

   ....

   ....

   ....

   ....

2. Suggest ways that the government could try to encourage people to recycle more.

   ....

   ....

   ....

Quick quiz

Are these statements **true (T)** or **false (F)**?

1. Plastics are a form of solid waste.   T   F
2. Metal cannot be recycled.   T   F
3. Donating clothes is one way of reducing solid waste.   T   F
4. Plastics take a year to breakdown.   T   F

Term 2 Unit 1 Materials and their properties

Self-check

Learning objectives	☺	😐	☹
I can identify different materials.			
I can describe proper storage of materials.			
I can identify properties of different materials.			
I can classify materials into groups using their properties.			
I can analyse properties of materials.			
I can link products with their properties.			
I can identify a range of objects and their uses.			
I can identify the best object for a use dependent on their properties.			
I can describe the ways that materials are disposed of.			
I can explain the impacts of different methods of disposal.			
I can identify materials that can be recycled at home.			
I can describe the benefits of recycling.			

Solids, liquids and gases

> **Learning objectives**
> - Describe what is meant by the terms 'solids', 'liquids' and 'gases'.
> - Classify some objects as solids, liquids and gases.

Make a note

Nearly every material can be classified as a solid, liquid or a gas.

Match the term with a description.

Term
solids
liquids
gases

Description
can flow and fill the bottom of a container
can fill the container they are in
can hold their shape

Classifying solids, liquids and gases

Look at these objects.

Classify them as solids, liquids or gases.

sugar

....

coffee

....

water

....

Term 2 Unit 1 Materials and their properties

Objects

Make a note

Some objects are difficult to classify.

1 Would you call butter a solid or a liquid? ....

2 Give a reason for your answer. ....

Quick quiz

Look at this list of objects. Circle all the liquids.

oil   ice   milk   steam   mango juice   seawater   sand

Properties of solids, liquids and gases

> **Learning objectives**
> - Identify some properties of solids, liquids and gases.

Properties of solids, liquids and gases

Make a note

Solids, liquids and gases have different properties.

Tick (✓) the boxes to show the different properties of a solid, a liquid and a gas.

	Can flow	Can take the shape of their container	Can hold their own shape	Can be compressed	Can't be compressed
solid					
liquid					
gas					

Solids, liquids and gases

Look around the environment you are in.

Write down the names of any solids, liquids and gases that you can see.

Solids:..

Liquids:...

Gases:..

Term 2 Unit 1 Materials and their properties

Particles in solids, liquids and gases

Make a note

Solids, liquids and gases have different properties because of how their particles are arranged.

The diagrams show the particles in solids, liquids and gases.
Write the name of the state under the correct diagram of their particles.

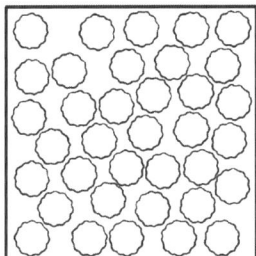   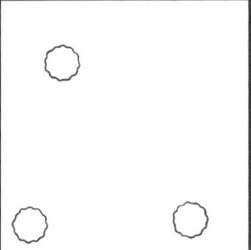   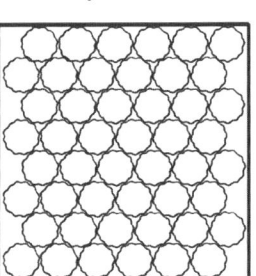

1 ....   2 ....   3 ....

Classifying objects

Make a note

Some objects are difficult to classify.

Describe a test you could do to see if toothpaste was a solid or a liquid.

....

Quick quiz

Are these statements **true (T)** or **false (F)**?
1 Liquids can be compressed.   T   F
2 Gases and liquids can flow.   T   F
3 The particles in a gas are closely packed together.   T   F
4 Solids take the shape of their container.   T   F

Changes of state

Learning objectives
- Describe how solids can be changed into liquids and gases.
- Use scientific words to describe the changes of state.

Make a note

Water can exist in three different states – solid, liquid and gas.

Match the state to the name of water in that state.

State
solid
liquid
gas

Name of water
steam
ice
water

Changing states

You are going to look a bit further at the changes of state of water by doing a series of mini experiments.

1 Get some ice cubes and heat them up in a pan on the stove.

Safety
Get an adult to help you.

Term 2 Unit 1 Materials and their properties

Circle the correct words in bold to show the change of state.

When we heat up ice, we change a **solid / liquid / gas** into a **solid / liquid / gas**.

This is called **melting / boiling / freezing**.

2   Put some water in a pan and heat it up on a stove until it bubbles.

Safety
Get an adult to help you.

Circle the correct words in bold to show the change of state.

When we heat up water, we change a **solid / liquid / gas** into a **solid / liquid / gas**.

This is called **melting / boiling / freezing**.

3   Put some water in a container in the freezer for a few hours and then have a look at it.

Circle the correct words in bold to show the change of state.

When we put water in a freezer, we change a **solid / liquid / gas** into a **solid / liquid / gas**.

This is called **melting / boiling / freezing**.

4   Put a mirror in the freezer for a few minutes and then take it out and breathe on in.

Make a note

The water in your breath hits the cold mirror and changes state.

Circle the correct words in bold to show the change of state.

When we breathe on a cold mirror, we change a **solid / liquid / gas** into a **solid / liquid / gas**.

This is called **melting / condensing / freezing**.

Changes of state

Describing changes of state

The diagram shows the changes in state of solids, liquids and gases.
Write the correct scientific names of the changes of state.
Use these words:

> boiling   condensing   melting   freezing

1 ....   2 ....

3 ....   4 ....

Quick quiz

Are these statements **true (T)** or **false (F)**?

1. Melting is when we change a liquid to a solid.   T   F
2. When we change a gas to a liquid, it is called 'condensation'.   T   F
3. Freezing changes a liquid to a solid.   T   F
4. When we change a solid to a liquid, it is called 'boiling'.   T   F

Term 2 Unit 1 Materials and their properties

Heating and cooling

> **Learning objectives**
> - Describe how temperature affects the changes of state.
> - Identify states using boiling and melting points.

Look around your home.

Identify a list of things that can be easily melted.

....

....

Heating curve data analysis task

A student heated ice until it boiled. They took the temperature and plotted their results on a graph.

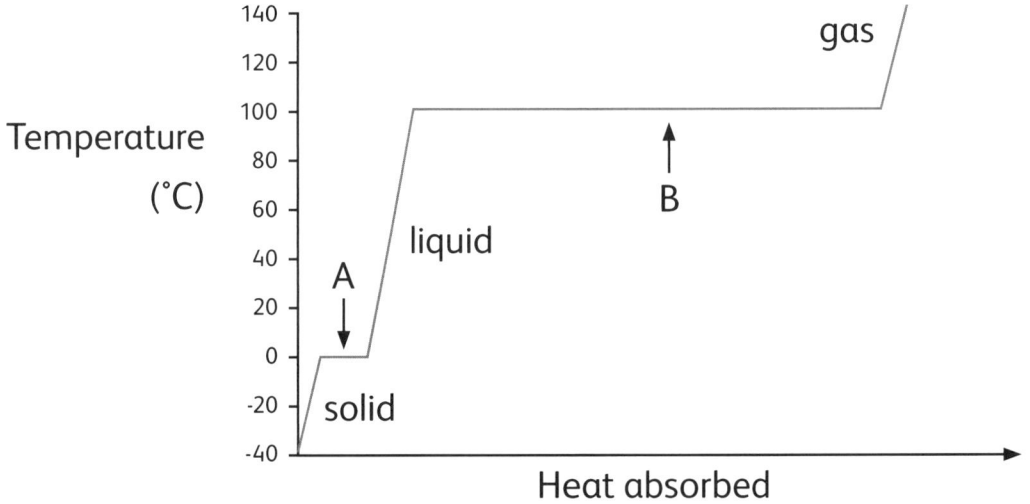

1. What temperature does water change from a solid to a liquid? .... °C

2. What temperature does water change from a liquid to a gas? .... °C

3. What term do we use to describe what is happening at A? ....

4. What term do we use to describe what is happening at B? ....

126

Heating and cooling

Changes of state in other substances

Make a note

The melting point is the temperature at which a solid changes into a liquid. The boiling point is the temperature at which a substance changes from a liquid to a gas.

ICT opportunity

Go online and research the melting and boiling points of the substances in the table.

1 Complete the table to show the melting point and boiling point of each substance and the state that the substance is at 20°C.

Substance	Melting point	Boiling point	State at 20°C
oxygen			
mercury			
lead			
benzene			

2 Which substance has the highest boiling point? ....

3 Which substance has the lowest melting point? ....

Quick quiz

Are these statements **true** (**T**) or **false** (**F**)?
1 The melting point is when a substance boils.   T   F
2 The boiling point is when a liquid changes to a gas. T   F
3 The melting point of water is 100°C.   T   F
4 All liquids boil at 100°C.   T   F

Term 2 Unit 1 Materials and their properties

Irreversible and reversible changes

> **Learning objectives**
> - Describe what is meant by irreversible and reversible changes.
> - Distinguish between irreversible and reversible changes.

ICT opportunity

Use a dictionary or go online to research the meaning of the words 'irreversible' and 'reversible'.

Write definitions.

irreversible:..

....

reversible: ....

....

Irreversible and reversible changes

Make a note

Irreversible changes can't be undone. Reversible changes mean that you can get the original substance back again.

Irreversible and reversible changes

Look at these changes. Can you classify them as irreversible or reversible?

Example of change	Irreversible or reversible?
melting	
burning	
freezing	
cooking	
boiling	

Irreversible and reversible changes in the kitchen

You will need:
- a saucepan
- a glass bowl
- some chocolate
- a frying pan
- an egg.

Safety
You might need an adult to help you.

ICT opportunity
If you can't do the experiment, then you can go online and read about solids, liquids, gases and changes of state.

Method

1   Melt some chocolate in a bowl over some boiling water.
    Describe what changes you see happening to the chocolate.

    ....

    ....

2   Take the chocolate off the heat and let it cool down.
    What do you notice about the chocolate? Write down your observations.

    ....

    ....

Term 2 Unit 1 Materials and their properties

3 Fry an egg in a frying pan.

Safety
Ask an adult to help you.

a) Describe what the egg looks like before you heat it.

....

....

b) What does the egg look like after you heat it? Write down your observations.

....

....

4 Cooking an egg is an irreversible change and melting chocolate is a reversible change.

a) Explain why we describe cooking eggs as an irreversible change.

....

....

b) Explain why we describe melting chocolate as a reversible change.

....

....

Quick quiz

Are these statements **true (T)** or **false (F)**?
1 Irreversible changes are permanent.   T   F
2 Melting is a reversible change.   T   F
3 Burning is a reversible change.   T   F
4 Boiling is a permanent change.   T   F

Examples of irreversible and reversible changes

> **Learning objectives**
> - Classify some changes as reversible and irreversible.

ICT opportunity

Watch a video about irreversible and reversible changes.

After watching the video, write down three irreversible and three reversible changes.

1 ....

2 ....

3 ....

4 ....

5 ....

6 ....

Irreversible and reversible changes

Look at these examples of changes. Can you classify them as irreversible or reversible?

ice melting   wood burning   a cake baking   a puddle evaporating

clay being moulded   sand and salt being mixed   metal rusting

Irreversible	Reversible

Term 2 Unit 1 Materials and their properties

Investigating a reversible change

> **Make a note**
>
> Some changes look like they are irreversible, but they are actually reversible.

You will need:
- some water
- some salt
- a glass
- a shallow saucer.

Method

1. Pour a small amount of water in a cup so that the cup is about one-third full.

2. Stir a spoonful of salt into the water until it dissolves. Keep doing this until no more salt will dissolve.

3. Pour the water into a shallow saucer and leave on the windowsill in the sun for a couple of days.

   a) What happens when you stir salt into the water?

   ....

   b) What can you see on the saucer after a couple of days?

   ....

   c) Is this an example of a reversible or irreversible change? Give a reason for your answer.

   ....

   ....

Irreversible and reversible changes quiz

ICT opportunity

Go online and find a quiz about irreversible and reversible changes.

Quick quiz

1 Name one irreversible change. ....
2 Name one reversible change. ....
3 Describe the difference between irreversible and reversible changes.

....

....

....

....

Term 2 Unit 1 Materials and their properties

Self-check

Learning objectives	😊	😐	☹
I can describe what is meant by the terms 'solids', 'liquids' and 'gases'.			
I can classify some objects as solids, liquids and gases.			
I can identify some properties of solids, liquids and gases.			
I can describe how solids can be changed into liquids and gases.			
I can use scientific words to describe the changes of state.			
I can describe how temperature affects the changes of state.			
I can identify states using boiling and melting points.			
I can describe what is meant by irreversible and reversible changes.			
I can distinguish between irreversible and reversible changes.			
I can classify some changes as reversible and irreversible.			

Extension activity

Aim: To explore the production and uses of the material glass.

Task: Research how glass is made and its uses to produce a factsheet.

ICT opportunity

Go online and research the task, using ideas that you have learned during this lesson. You can make notes on the next page.

Design your factsheet on page 137.

Things to include:
1. What glass is made from.
2. What the properties of glass are.
3. What the uses of glass are.
4. How glass is made.
5. Whether the production of glass is a reversible or irreversible change.

Term 2 Unit 1 Materials and their properties

Notes

Extension activity

Factsheet

Term 2 Unit 1 Materials and their properties

Practice test

The table shows the properties of some different materials. Use the table to help you answer questions 1, 2 and 3.

Material	Can conduct heat	Can conduct electricity	Is transparent	Is an insulator	Is flexible	Is hard
A	✓	✓			✓	
B				✓	✓	
C			✓	✓		
D				✓		✓

1 Which material would be best for making warm clothing? Circle.

   A   B   C   D

2 Which material would be best for making electrical wires? Circle.

   A   B   C   D

3 Which material is glass? Circle.

   A   B   C   D

Here is a list of properties that different materials have.

A is hard

B is a conductor

C is waterproof

D is resistant to high temperatures

E is flexible

4 Which properties would a material that is used to make a saucepan need?

....

5 Which properties would a material that is used to make clothing for water sports need?

....

This graph shows some data about the percentage of different materials that are recycled.

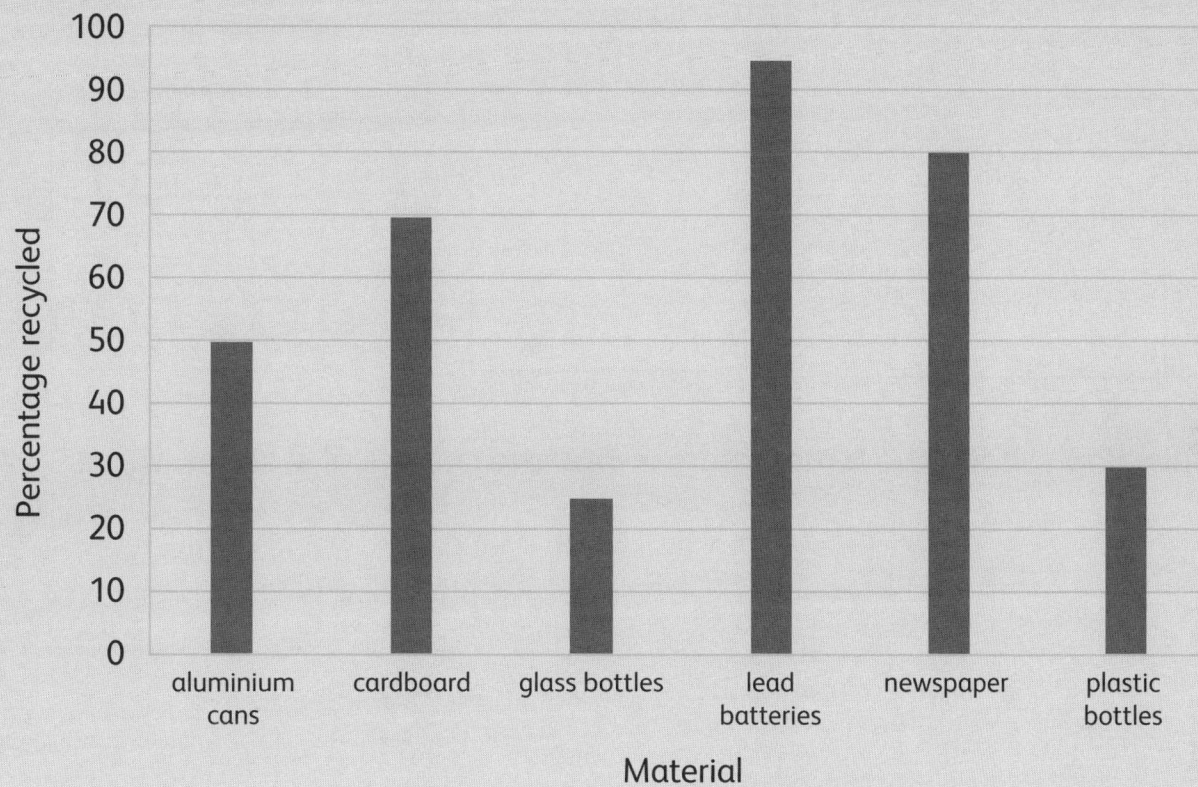

6 Name the most commonly recycled material.

....

7 What percentage of newspaper is recycled?

....

8 What are the benefits of recycling aluminium cans? You can circle more than one.

   a) conserves natural resources

   b) doesn't use any energy

   c) prevents solid waste pollution

   d) prevents air pollution

Term 2 Unit 1 Materials and their properties

This sentence describes a state of matter:

A substance can flow and this substance can be compressed and takes up the space of the container it is in.

9 What state of matter does this sentence describe? ....

10 What do the particles look like in this state of matter?

A 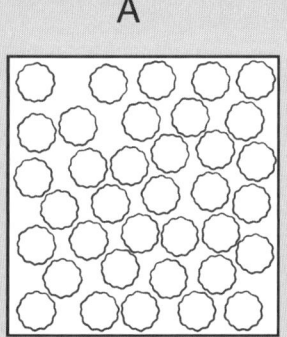 B 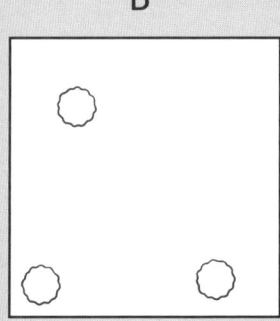 C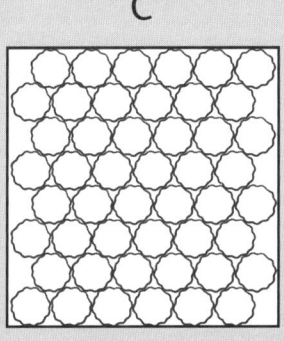

This graph shows the temperature during the change of state of water.

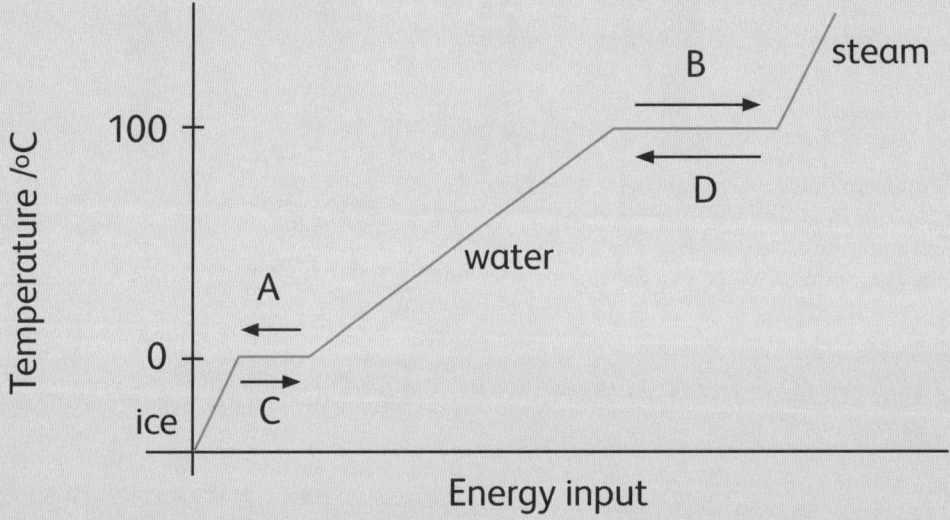

11 What change of state does the letter B represent? ....

12 Which letter represents freezing? ....

This table shows the melting and boiling point of five different substances.

Substance	Melting point / °C	Boiling point / °C
A	−5	101
B	−71	−61
C	961	2 162
D	−140	−1
E	−63	61

13 Which substances are gases at room temperature (20°C)?

....

14 Which substances are solid at 0°C?

....

15 Explain why melting and boiling are reversible changes.

....

....

Term 2 Unit 2 Human body systems

Human Body Systems

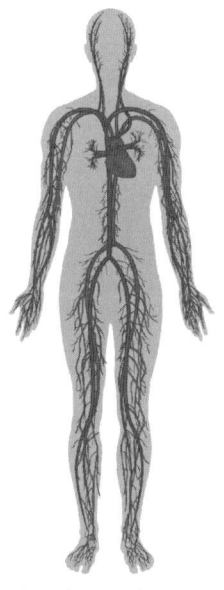
Circulatory System

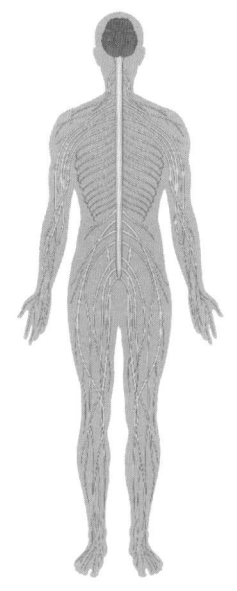
Nervous System

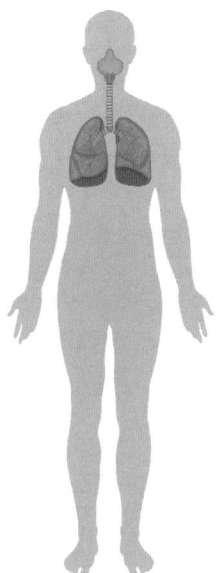
Respiratory System

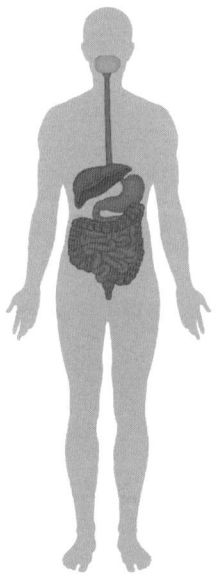
Digestive System

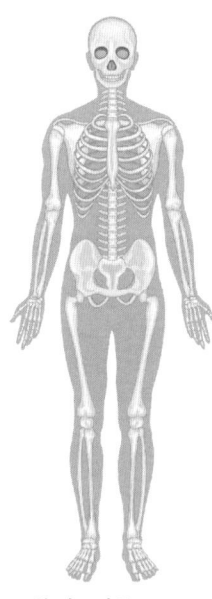
Skeletal System

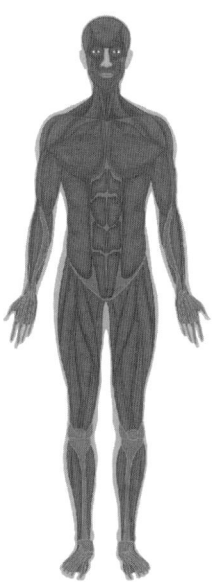
Muscular System

Different body systems

> **Learning objectives**
> - Explain what is meant by the term 'system'.
> - Identify some organ systems and their functions.

ICT opportunity

Look up the term 'system' in a dictionary or on the internet.

Write down what 'system' means. ....

....

Can you think of any organ systems in the body?
Write down the names of any that you can think of.

....

....

....

In this lesson, you will learn about human body systems. You will look at the parts of different human body systems and their functions.

Match the body systems

Make a note

Humans have different body systems containing different organs that have specialised functions in the body.

Term 2 Unit 2 Human body systems

Look at the pictures of the different body systems.

Write the correct name of each body system under its picture, using the systems that follow.

circulatory system   **digestive system**   **muscular system**

male and female reproductive system   **respiratory system**

skeletal system

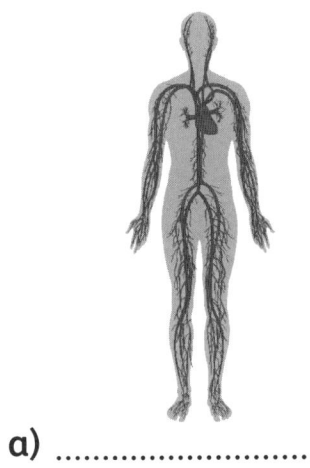

a) ....

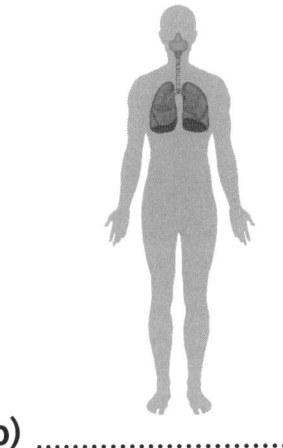

b) ....

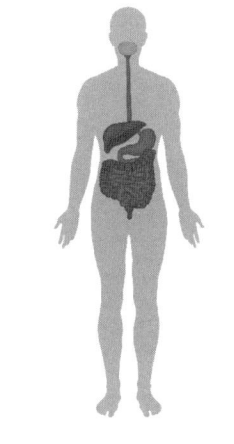

c) ....

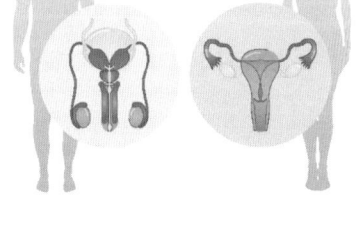

d) ....

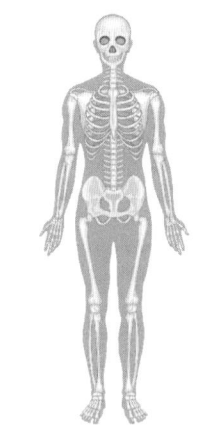

e) ....

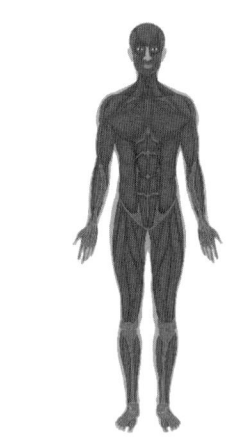

f) ....

Functions of the different body systems

The boxes on the left show some body systems. The boxes on the right show the functions of the systems.

Draw lines to link the body system with its function.

Body system
circulatory system
digestive system
respiratory system
skeletal system

Function
to move
to transfer blood around the body
to breakdown food and absorb nutrients
to breathe

Quick quiz

Are these statements **true (T)** or **false (F)**?

1. The heart is part of circulatory system. T F
2. The heart is part of the respiratory system. T F
3. You can't move without a skeleton. T F
4. The stomach is part of the digestive system. T F

Term 2 Unit 2 Human body systems

The circulatory system

Learning objectives
- Identify the parts of the circulatory system.

Place your two fingers on your wrists like this.
See if you can feel your pulse.

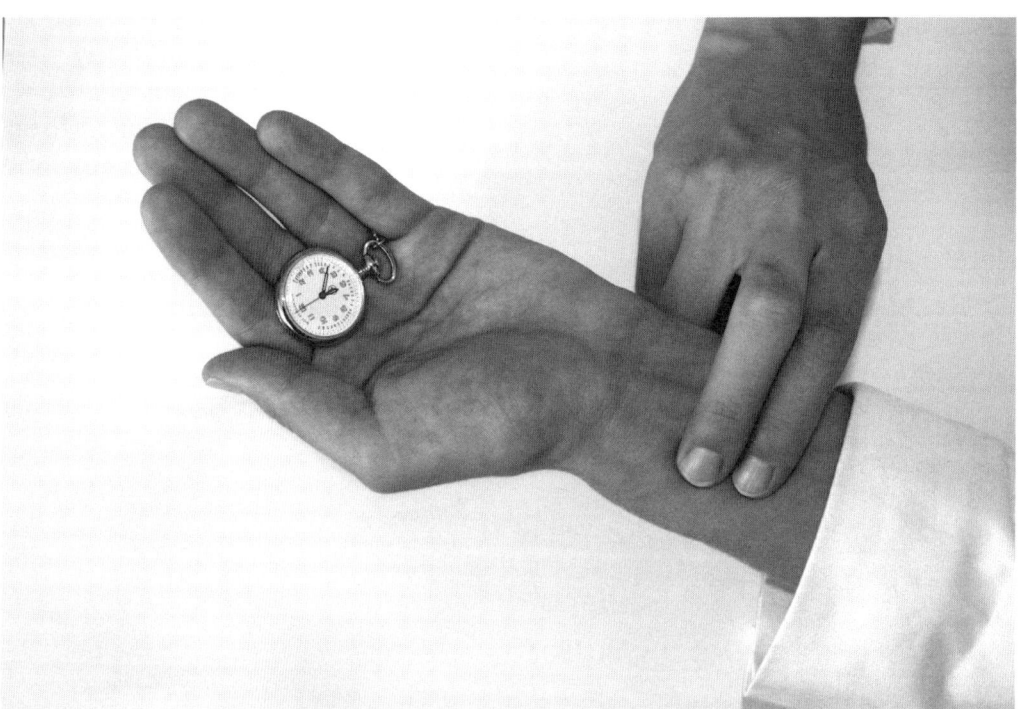

Make a note

The pulse you can feel is the pulse of blood flowing close to the surface of the skin. It is caused by the heart beating and pushing blood around the body.

See if you can find your pulse rate in one minute.

In this lesson, you will look at the different parts of the circulatory system.

The circulatory system

Parts of the circulatory system

Make a note

The role of the circulatory system is to transport blood around the body.

The diagram shows a simplified circulatory system.

Label the diagram to show the parts of the circulatory system using the words that follow.

lungs   heart   body   pulmonary artery
pulmonary vein   arteries   veins

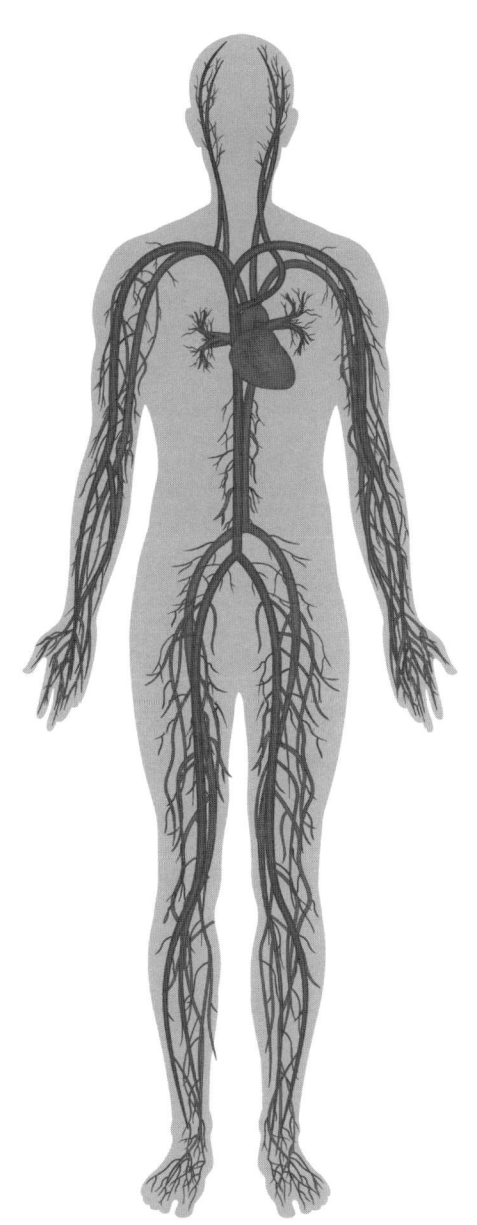

Term 2 Unit 2 Human body systems

The respiratory system

> **Learning objectives**
> - Identify the parts and functions of the respiratory system.

Place your hands on your chest.

Take a deep breath in and out. What do you notice happening to your chest as you breathe in and out?

....

....

Time yourself for one minute. Count how many breaths you take in a minute. Write down your answer.

.... breaths per minute

The respiratory system

Make a note

The respiratory system is the body system that has to do with breathing.

The diagram on the next page shows the respiratory system.

ICT opportunity

Go online and research the parts of the respiratory system.

The respiratory system

Label all the parts of the respiratory system you can find.

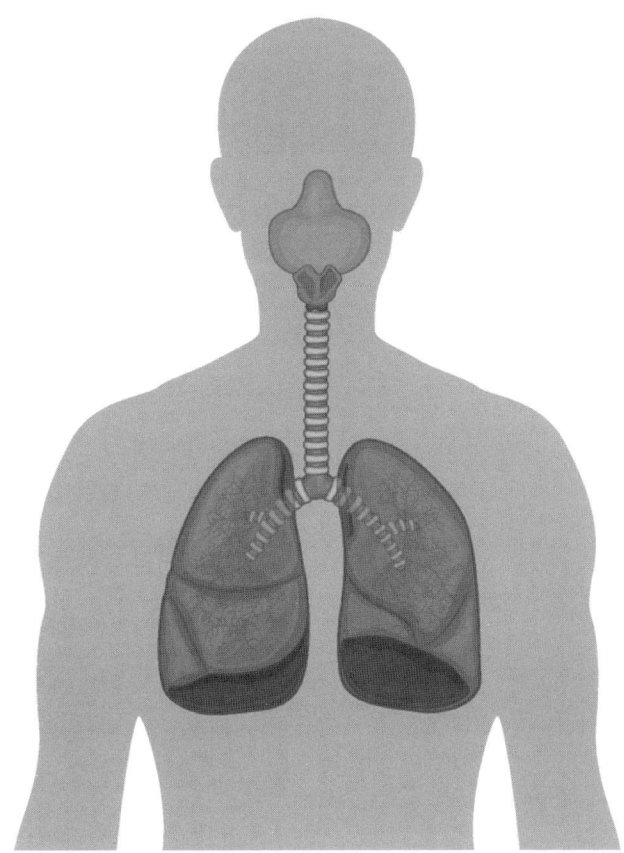

The role of the respiratory system

ICT opportunity

Watch a video about the respiratory system.

Hint

It has a large surface area.

Use the information and your knowledge to answer the following questions.

1 What gas do we need from the air that we breathe? ....

2 What gas is a waste gas that we breathe out? ....

3 What is the name of the part where oxygen is absorbed?

....

....

Term 2 Unit 2 Human body systems

4 Describe what happens when you hold your breath. ....

....

The respiratory system and exercise

1. Write down how many breaths per minute you got for your starter activity here.

   .... breaths per minute

> **Make a note**
>
> The number of breaths you take per minute is your resting breathing rate.

2. Run on the spot for two minutes.

   Write down how many breaths per minute you have now.

   .... breaths per minute

> **! Safety**
>
> Make sure you have enough space and you are not going to trip over anything.

3. Calculate the difference between your resting breathing rate and your breathing rate after exercise.

   .... breaths per minute

4. What do you notice about your breathing rate when you exercise?

   ....

> **Quick quiz**
>
> Are these statements **true** (T) or **false** (F)?
> 1. Carbon dioxide is the gas we need when we breathe.   T   F
> 2. Breathing rate increases with exercise.   T   F
> 3. When you breathe in, air goes down your trachea.   T   F

The male reproductive system

> **Learning objectives**
> - Identify the parts of the male reproductive system.

Use the letters from the term 'male reproductive system' to make as many new words as you can.

Write down the words here. How many can you make?

....

....

....

In this lesson, you will learn about the parts of the male reproductive system.

The parts of the male reproductive system

Label the parts of the male reproductive system using the labels that follow.

      penis   **scrotum**   **sperm duct**   **testes**   **urethra**

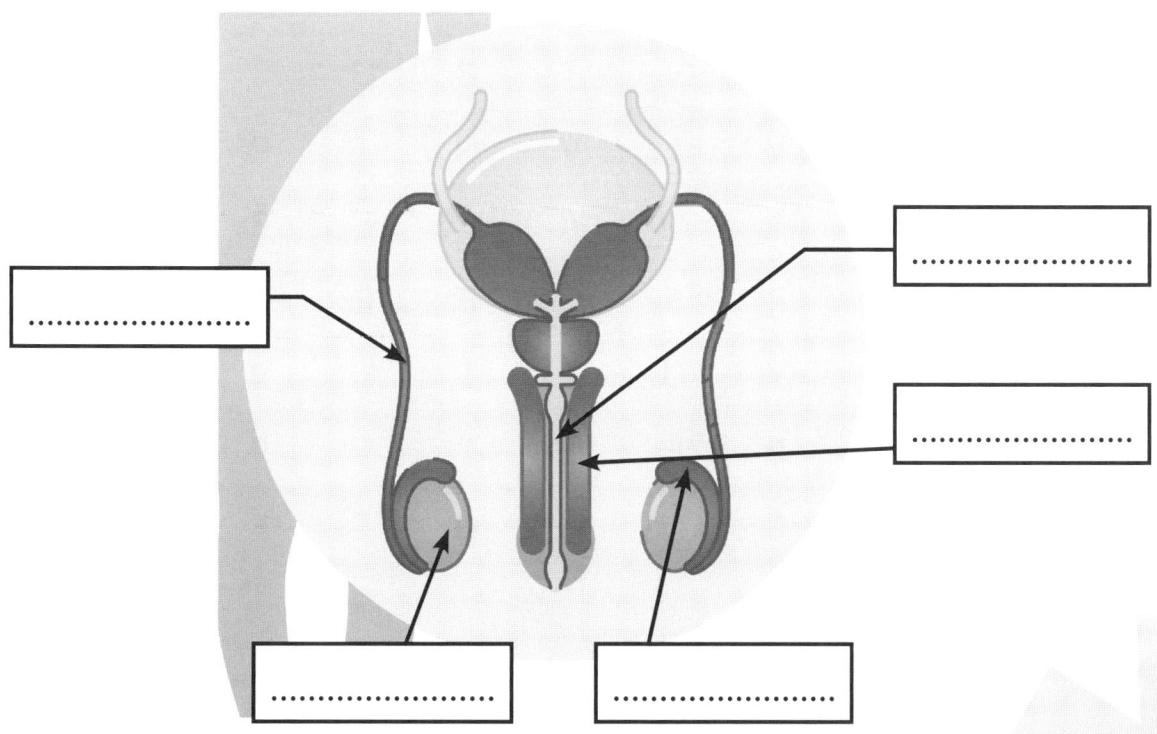

Term 2 Unit 2 Human body systems

The female reproductive system

Learning objectives
- Identify the parts of the female reproductive system.

Make a note

The gestation period is how long an animal is pregnant for.

ICT opportunity

Go online and research how long the following animals are pregnant for.
1. humans ....
2. giraffe ....
3. elephant ....
4. mongoose ....

In this lesson, you will learn about the parts of the female reproductive system.

The parts of the female reproductive system

Find these words in the wordsearch.

oviduct
ovary
uterus
vagina

D	F	S	T	O	V	D
O	V	I	D	U	C	T
V	L	V	N	T	P	E
A	M	A	C	E	S	S
R	S	G	E	R	T	T
Y	U	I	T	U	V	E
P	E	N	I	S	P	S
G	N	A	O	T	E	T

The digestive system 1

> **Learning objectives**
> - Identify the parts and functions of the digestive system.

Think about what happens when you eat your lunch.

Where does your food go? Can you write down any names of the parts the food passes through?

Write down your ideas.

....

....

....

In this lesson, you will learn about the parts and function of the digestive system.

The parts of the digestive system

Make a note

The role of the digestive system is to break down food into nutrients we can use in the body.

Term 2 Unit 2 Human body systems

Use these words to label the digestive system diagram.

mouth   oesophagus   stomach   small intestine
large intestine   pancreas   anus

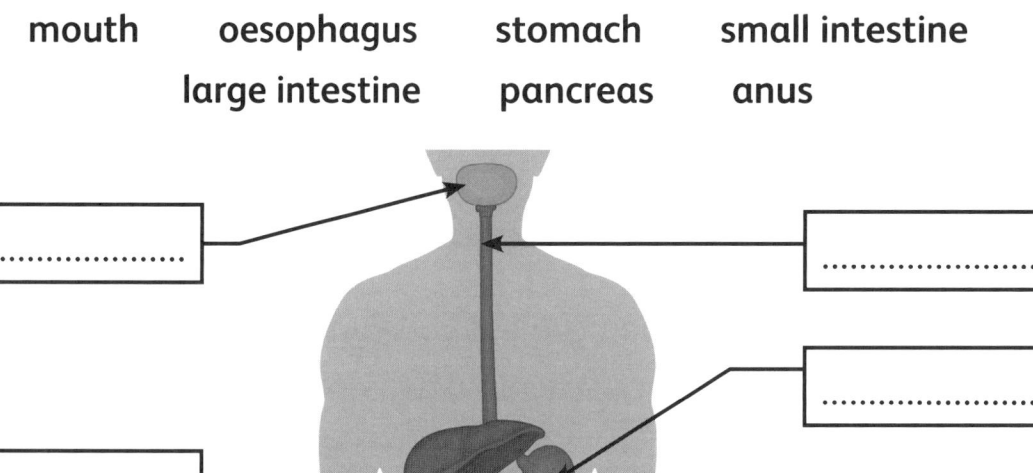

The functions of the parts of the digestive system

The boxes on the left show parts of the digestive system. The boxes on the right show the functions of these parts.

Draw lines to link the part of the digestive system with its function.

Part	Function
anus	physically breaks down food into smaller pieces
large intestine	transfers food to the stomach
mouth	produces digestive enzymes
oesophagus	absorbs most of the nutrients from food
pancreas	absorbs excess water
small intestine	contains acid to kill bacteria
stomach	releases waste from the body

The digestive system 1

The liver

Make a note

The liver is an important organ and has over 300 functions.

ICT opportunity

Go online and read information about the liver.

Write down three main functions of the liver.

1 ....

2 ....

3 ....

Quick quiz

1 Where is food taken into the digestive system? ....

2 Which part of the digestive system kills bacteria? ....

3 Where are the nutrients from food absorbed? ....

Term 2 Unit 2 Human body systems

The digestive system 2

Learning objectives
- Build a model digestive system.
- Describe the pathway of food through the digestive system.

Check if you can remember the names of all the parts of the digestive system.

Write them down here:

....

....

In this lesson, you will build a model digestive system to find out how it works and describe the pathway of some food through the digestive system.

Building a model digestive system

You will need:
- a small towel
- a glass of water
- a glass of orange juice
- a banana
- a bowl
- a sealable plastic bag
- some tights
- some cookies
- kitchen paper
- scissors.

Method

1. Put the water, orange juice, banana and cookies in the plastic bag and seal it.

2. Churn the food by squeezing the bag until all the contents are crushed up.

Safety
Take care when using scissors.

The digestive system 2

3 Transfer the contents to the tights.

4 Squeeze out as much of the liquid as you can into the bowl.

5 Dry the outside of the tights with your towel.

6 Cut a small hole in the bottom of the tights.

7 Squeeze the mashed food out of the tights. This represents your poo!

8 Think about what part of the digestive system each part of the method represents.

Part 2: churning of the food ....

Part 4: squeezing the liquid out of the tights ....

Part 6: drying the tights with a towel ....

The story of a burger

Imagine that you are a burger that is being eaten.

Write about your pathway through the digestive system.

Think about how you would feel. What different things would you experience as you go through the digestive system. Try to use all the correct scientific words.

Include the following words in your story:

mouth   teeth   stomach   small intestine   large intestine

oesophagus   saliva   enzymes   water   anus

....
....
....
....
....
....

Term 2 Unit 2 Human body systems

Quick quiz

1 Name two functions of the stomach.

....

....

2 Name where food enters the digestive system.   ....

3 Name where waste comes out of the digestive system.   ....

The skeletal system

Learning objectives
- Describe some of the major bones in the skeletal system.
- Describe some of the functions of the skeleton.

Make a note

The skeleton consists of many bones and has several important functions.

ICT opportunity

Watch a video about our bones.

Write down the names of three bones that you have not heard of before and write down where you find them.

1 ....

2 ....

3 ....

In this lesson, you will look at the main bones of the skeletal system and the important functions of the skeleton.

The skeleton

Make a note

There are over 200 bones in the human body.

Term 2 Unit 2 Human body systems

Look at this diagram of a skeleton.

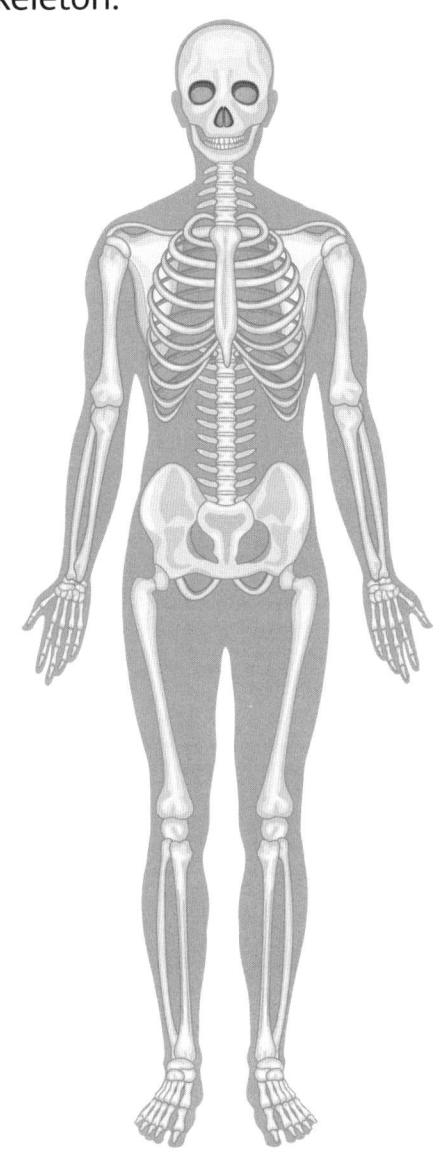

Use the diagram of a skeleton to help you answer these questions.

1 Name the three bones of the arm.

   a) .... b) .... c) ....

2 What is the proper name for your knee bone?

   ....

3 Name the three bones of the leg.

   a) .... b) .... c) ....

4 What are your toe bones and finger bones called? ....

5 Name the bone that protects your brain. ....

The skeletal system

The skeleton web activity

ICT opportunity

Go online and read information about the skeletal system.

Use the information and your knowledge to answer these questions.

1 Where do you find the smallest bone in your body? ....

2 Name two functions of the skeletal system.

   a) .... b) ....

3 How many bones are in your spine? ....

4 What important job does your bone marrow do?

   ....

5 What are 70% of your bones made from? ....

6 What do we call the places where bones meet? ....

Quick quiz

Are these statements **true (T)** or **false (F)**?

1 The skeletal system is only made of dead tissue.   T   F
2 The human body has 100 bones.   T   F
3 Bone marrow produces red blood cells.   T   F
4 The skeleton is important for movement.   T   F

Term 2 Unit 2 Human body systems

Muscles and joints

> **Learning objectives**
> - Identify some major muscle and joint types.
> - Describe how the bones, joints and muscles work together to produce movement in humans.

Think about how your body moves.

Make a note

Your muscles, joints and muscles work together to help your body move.

Write down the names of as many muscles as you can think of.

....

....

In this lesson, you will learn about the major muscle groups and how bones, joints and muscles work together to help you move.

Major muscle groups

1 Match the name of the muscle to where you would find it in the body.

Muscle
bicep
pectorals
trapezius
obliques
quadriceps

Where you find the muscle in the body
chest
back
abdominal
legs
arm

Muscles and joints

2 Can you label any of the muscles on this diagram?

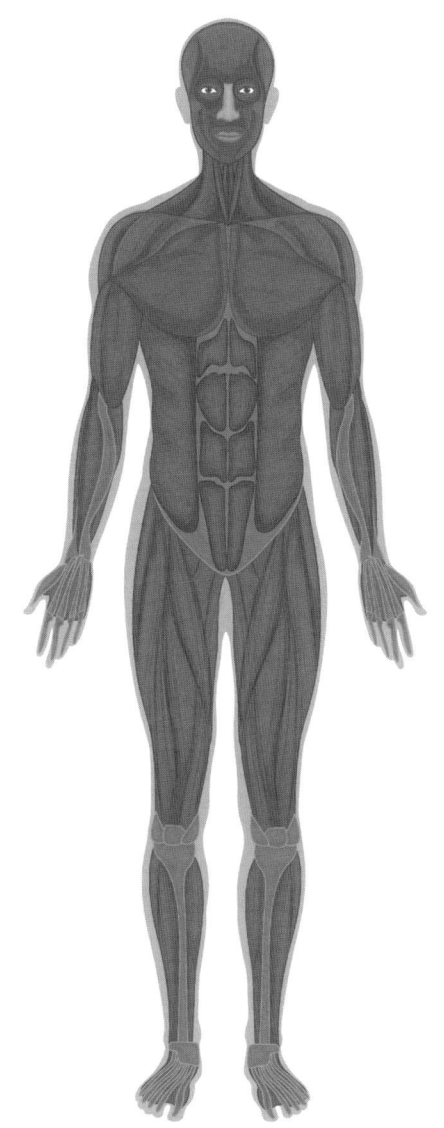

Types of joints

Make a note

The place where two bones meet in the body is called a 'joint'. There are many different types of joints in the body.

The diagrams below show two different types of joints.

    A ball and socket joint   **B** hinge joint

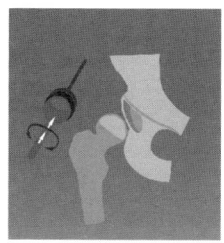 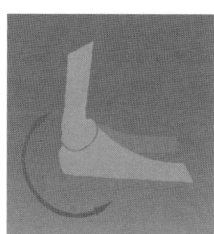

Term 2 Unit 2 Human body systems

Make a note

Your shoulder is an example of ball and socket joint. Your elbow is an example of a hinge joint.

1. Move your shoulder. Note which directions you are able to move your shoulder. Write down your observations.

   ....

2. Move your elbow. Note which directions you are able to move your elbow. Write down your observations.

   ....

3. Write down another example of a ball and socket joint. ....

4. Write down another example of a hinge joint. ....

5. Can you find out the names of any other types of joints? Write them down.

   ....

Quick quiz

Are these statements **true** (T) or **false** (F)?
1. You need bones to help you move.   T   F
2. Joints are where bones meet muscles.   T   F
3. Your elbow is an example of a ball and socket joint.   T   F
4. Your hamstring is an example of a muscle in your arm.   T   F

Investigating movement

Learning objectives
- Describe how the bones, joints and muscles work together to produce movement in humans.
- Investigate how the elbow works.

ICT opportunity

Watch a video about the warm-up exercises that police recruits have to do.

See if you can join in with the exercises.

Safety

Make sure you have enough space around you to do this safely.

Think about all the bones, muscles and joints in your body moving to help you do this.

See if you can count the number of different exercises there are.

In this lesson, you will learn about how bones, joints and muscles work together to help you move and investigate the movement of your elbow.

Connective tissues

Make a note

Ligaments connect bones to bones. Tendons connect bones to muscles.

Term 2 Unit 2 Human body systems

ICT opportunity

Go online and read information about the position of these connective tissues to help you find out if it is a tendon or a ligament.

1 Complete the table.

What does it connect?	Tendon or ligament?
patella and tibia	
hamstring and heel bone	
trapezius and scapula	
pelvis and femur	
pectorals and sternum	

ICT opportunity

Can you find out any names of major ligaments and tendons?

2 Write them in the space below.

ligaments: ....

tendons: ....

Making a model elbow

Safety
Take care when using a craft knife.
Ask an adult to help you.

You will need:
- a card
- a ruler
- a craft knife
- a masking tape
- a large paperclip
- long balloons.

Investigating movement

1 Make a model arm using the instructions.

Instructions	Diagram
1 Cut out two pieces of card measuring 20 cm x 28 cm.	1–4
2 Roll them up into cylinders to look like bones and secure them with masking tape. Write the words 'radius' on one and 'ulna' on the other.	
3 Cut out another piece of card measuring 28 cm x 30 cm.	5–6
4 Roll it up and secure with masking tape. Write the word 'humerus' on it.	
5 Put the 'bones' in the arrangement you can see in the picture. Using a craft knife, make a hole through each bit of card as shown.	7–8
6 Straighten a paper clip and thread it through the hole. Bend the ends to form hooks so the paperclip doesn't come out of the hole.	
7 Partially inflate two long balloons. These will be the muscles. Write 'bicep' on one balloon and 'tricep' on the other balloon.	9
8 Tie one end of the bicep to the radius and ulna near the elbow joint and the other end to the top of the humerus.	
9 Tie one end of the tricep to the radius and ulna near the elbow joint and the other end round the back of the elbow to the top of the humerus.	

Term 2 Unit 2 Human body systems

2 Answer these questions using your model arm.

a) Bend the elbow. See which muscle gets bigger (contracts) and which muscle stretches out (relaxes). Write these down.

....

b) Straighten the elbow. See which muscle gets bigger (contracts) and which muscle stretches out (relaxes). Write these down.

....

Make a note

These muscles work in opposing pairs.

3 Find out the name for these types of muscles. ....

4 What is missing from your model? ....

Quick quiz

Complete these sentences about bones, muscles and tendons.

1 Ligaments connect bone to ....

2 Tendons connect bone to ....

3 Some muscles work is .... pairs.

4 When one muscle contracts, the other ....

Excretion

> **Learning objectives**
> - Identify the excretory organs of humans.
> - State the role of organs in excretion.

Think about and write down the way water enters your body.

....

Think about and write down ways water leaves the body.

....

....

In this lesson, you will identify the different excretory organs of the body and describe what they do.

The urinary system

Make a note

The urinary system is responsible for removing urea, excess water and excess salts in the form of urine from the body.

Term 2 Unit 2 Human body systems

1 Label the urinary system with the following words:

kidney   bladder   ureter

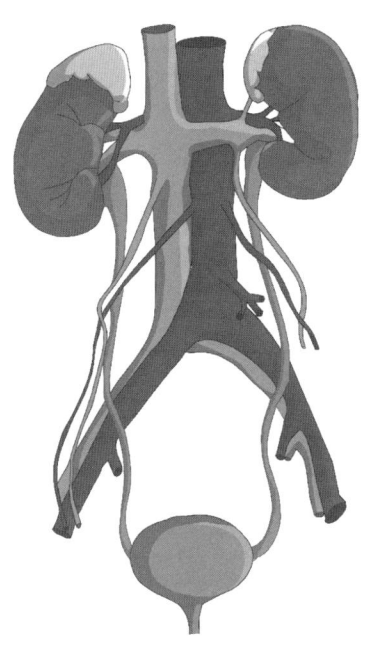

ICT opportunity

Listen to a podcast about the urinary system.

2 Write down the four jobs of the urinary system.

....

....

....

....

Excretion

Excretion activity

Read the information in the text box and use it to help you to answer the questions.

> The excretory system is the system of an organism's body that performs the function of excretion, the bodily process of getting rid of waste products.
>
> There are several parts of the body that are involved in this process, such as sweat glands, the liver, the lungs and the kidney system.
>
> Urea is a toxic waste product made in the liver. It is transported to the kidneys where it is mixed with excess water and salts to form urine. It is then transported to the bladder and released from the body.
>
> Your lungs also have a role in excretion to get rid of the toxic gas carbon dioxide. You breathe it out.

1 Name two waste products that the body needs to get rid of.

a) .... b) ....

2 What is urine made from?

a) .... b) .... c) ....

3 Name the organ that stores urine until it is ready to be released.

....

4 Name the organ that is used to remove carbon dioxide from the body.

....

Quick quiz

Are these statements **true (T)** or **false (F)**?

1 Sweat is one way in which water is released from the body. T   F
2 Oxygen is a waste product.   T   F
3 The kidneys make urea.   T   F

Term 2 Unit 2 Human body systems

The nervous system

> **Learning objectives**
> - Describe the parts of the nervous system.
> - Complete an investigation to test reflexes.

Think about what happens when you touch something very hot. How did you react? Did you have to think about your reaction? Write down your thoughts here.

....

....

Make a note

This type of reaction is called a 'reflex reaction' and it is controlled by your nervous system.

Can you think of another reflex reaction? ....

In this lesson, you will identify the parts of the nervous system and also carry out an investigation to test your own reflexes.

The nervous system web activity

ICT opportunity

Go online and read information about the nervous system.

The nervous system

Use the information to answer these questions.

1 How fast can your body send messages to the brain? ....

2 Name the three parts of your nervous system.

   a) .... b) .... c) ....

3 What is the proper name for nerve cells? ....

4 What are the two parts of the central nervous system (CNS)?

   a) .... b) ....

5 What is the peripheral nervous system made of? ....

Testing your reactions

ICT opportunity

Find an online game to test your reaction time. You can try this a few times.

Write down your fastest time here. ....

Testing your reactions investigation

You will need a friend for this activity.

1 Collect a ruler and a friend.

2 Ask your friend to hold a ruler up and you put your index finger and thumb apart at the bottom of the ruler. Make sure '0 cm' is at the bottom.

3 Your friend will let go of the ruler without warning.

4 Catch the ruler as quickly as you can and read how many centimetres it took for you to catch the ruler.

5 Get your friend to do the same and write your results in the table on the next page.

Term 2 Unit 2 Human body systems

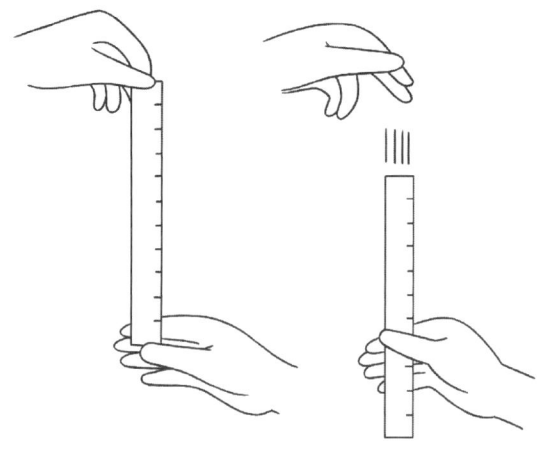

	Reaction time/cm	
	You	Your friend
Test 1		
Test 2		
Test 3		
Average		

6 Who had the fastest reaction time – you or your friend? ....

7 Think about the ways that you could make this investigation a fair test. Write down your ideas.

....

....

Quick quiz

Are these statements **true** (T) or **false** (F)?
1 The brain is part of the peripheral nervous system.   T   F
2 Neurones are nerve cells.   T   F
3 A reflex reaction requires you to think about it.   T   F

Self-check

Learning objectives	😊	😐	☹
I can explain what is meant by the term 'system'.			
I can identify some organ systems and their functions.			
I can identify the parts of the circulatory system.			
I can identify the parts and functions of the respiratory system.			
I can identify the parts of the male reproductive system.			
I can identify the parts of the female reproductive system.			
I can identify the parts and functions of the digestive system.			
I can build a model digestive system.			
I can describe the pathway of food through the digestive system.			
I can describe some of the major bones in the skeletal system.			
I can describe some of the functions of the skeleton.			
I can identify some major muscle and joint types.			
I can describe how the bones, joints and muscles work together to produce movement in humans.			
I can investigate how the elbow works.			
I can identify the excretory organs of humans.			
I can state the role of organs in excretion.			
I can describe the parts of the nervous system.			
I can complete an investigation to test reflexes.			

Term 2 Unit 2 Human body systems

Extension activity

Aim: To review the major body systems and their functions.

Task: To create a mind map of the different body systems studied during this topic.

Look up images of mind maps. Your task is to produce a mind map for this topic.

Include diagrams and colour to make your mind map more interesting.

Things to include:

1 The names of the different body systems.

2 The organs or parts of the different body systems.

3 The function of the different body systems.

4 Diagrams of the different body systems. You can draw these or print out a picture and stick it in.

5 Make notes on the opposite page and draw your mind map on page 178.

Extension activity

Notes

Term 2 Unit 2 Human body systems

Mind map

Practice test

Look at the table of a person's pulse rate doing different activities.

Activity	Heart rate / beats per minute
resting	72
walking	102
running	154
cycling	125
playing football	133

1. Which activity results in the highest heart rate? ....

2. Calculate the increase in heart rate when you go from resting to cycling.

   .... beats per minute

3. When you exercise, your breathing rate increases. Which statements explain why? You can circle more than one.

   a) You need to breathe in more carbon dioxide.

   b) You need to breathe in more oxygen.

   c) You need to breathe out more carbon dioxide.

   d) You need to breathe out more oxygen.

4. This table shows some organs. Tick (✓) which organs belong to which body system.

Body system	Brain	Heart	Stomach	Bladder
nervous				
excretory				

This diagram shows a simplified version of the circulatory system.

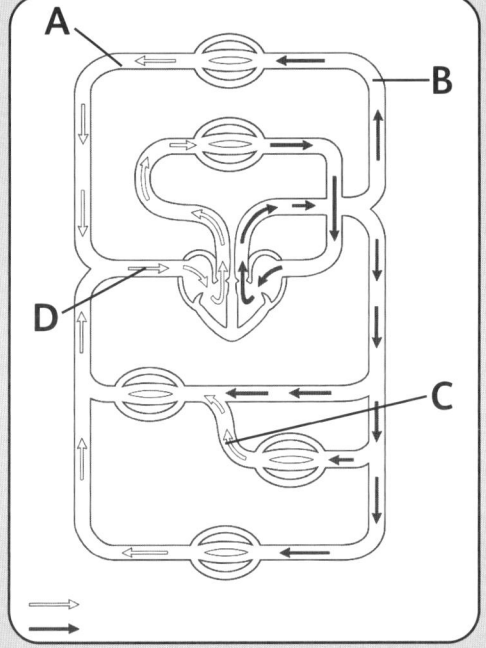

5 Which letter represents blood that carries the most oxygen? Circle.

   A   B   C   D

6 Which letter represents an artery? Circle.

   A   B   C   D

7 Complete the flow chart to show the pathway of urine out of the body.

kidney → .... → **bladder** → ....

8 Think about when you bend your arm.

Tick (✓) which muscles contract and relax.

Muscle	Contract	Relax
bicep		
tricep		

9 Which are reasons that you do not have to think about a reflex action? You can circle more than one.

   a) because the nervous system is not involved

   b) so that you can react more quickly

   c) so you don't need to include the brain

   d) so the body can automatically protect itself

10 Which is the correct pathway through the digestive system? Circle the correct answer.
   a) mouth → large intestine → small intestine → stomach → oesophagus
   b) mouth → small intestine → stomach → oesophagus → large intestine
   c) mouth → oesophagus → stomach → large intestine → small intestine
   d) mouth → oesophagus → stomach → small intestine → large intestine

Acid pH values are from 1–6. Neutral pH is 7. Alkali pH values are 8–12. Look at the table of different pH values inside different organs in the digestive system.

Organ	pH
A	7
B	10
C	2
D	8

11 Which organ is most likely to be the stomach? Circle.
   A   B   C   D

12 Which of these are excretory products? You can circle more than one.
   a) carbon dioxide
   b) oxygen
   c) urea
   d) proteins
   e) excess water

Term 2 Unit 2 Human body systems

This diagram shows a knee joint.

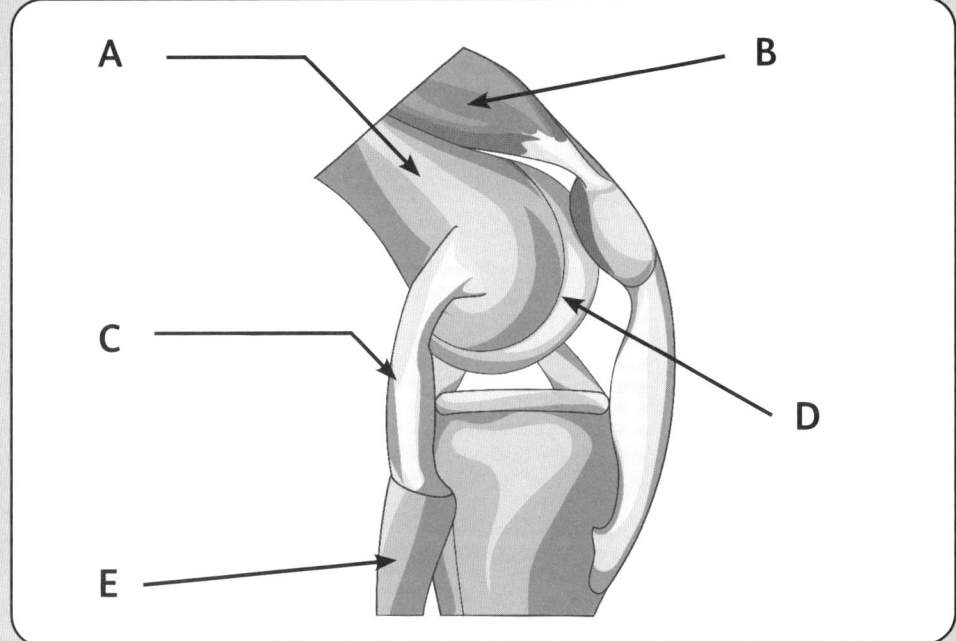

13 Name this type of joint. ....

14 Which letter represents a ligament? Circle.

    A   B   C   D   E

15 Name the bone labelled E. ....

Term 3 Unit 1 — Mixtures

What is a mixture?

Learning objectives
- Demonstrate that a mixture is made up of two or more substances.
- Recognise that all mixtures can be separated.

Make some hibiscus iced tea by following the instructions.
If you have a family recipe you can use that instead.

Safety
Take care not to burn yourself with hot water. Ask an adult to help you.

You will need:

- 15 g dried hibiscus flowers
- a stick of cinnamon
- 1 litre of hot water
- 1 lime, cut in quarters
- 2 tablespoons of sugar.

Method
1. Mix the hibiscus flowers, cinammon and water in a large jug.
2. Cool for an hour or two, then put the jug in a fridge overnight.
3. Strain the liquid to remove the solid pieces. Add ice, lime wedges and sugar to taste. Enjoy!

Make a note

When you mix ingredients together and no irreversible change happens, a mixture is made.

What is a mixture?

In this lesson, you will learn about mixtures. You will look at what a mixture is and find out about separating mixtures.

Mixture wordsearch

Find these words in the wordsearch. These are some of the words we will use in this unit.

| mixture | suspension | filtration | decanting |
| colloid | solution | evaporation | sieving |

A	T	I	O	C	N	E	S	A	I	O	N	S
E	V	A	P	O	R	A	T	I	O	N	M	I
M	E	I	V	L	T	O	O	N	M	A	L	E
I	A	F	I	L	T	R	A	T	I	O	N	V
X	M	N	S	O	L	U	T	I	O	N	D	I
T	S	I	S	I	U	O	I	N	L	R	C	N
U	T	S	C	D	E	C	A	N	T	I	N	G
R	S	U	S	P	E	N	S	I	O	N	M	I
E	T	I	S	O	M	X	T	S	U	S	D	X

Term 3 Unit 1 Mixtures

Different mixtures

You made a mixture during the starter activity.

Make a note

Mixtures are combinations of two or more substances that are not chemically combined. You can separate mixtures using different separating techniques.

Complete the table to show whether you think the following are mixtures or not mixtures.

Substance	Mixture or not mixture?
sea water	
pure water	
air	
gold	
soil	

Quick quiz

Are these statements **true** (**T**) or **false** (**F**)?
1. Mixtures cannot be separated.   T   F
2. Filtration is a separation technique.   T   F
3. Making a mixture is an irreversible change.   T   F
4. Sand is a mixture.   T   F

Classifying mixtures

> **Learning objectives**
> - Classify mixtures as solutions, suspensions and colloids.

Match the type of mixture with its definition.

Type of mixture
A colloid
A solution
A suspension

Definition
is a transparent mixture.
is a cloudy mixture where the solid particles will easily separate out.
is a cloudy mixture where the solid particles will not easily separate out.

In this lesson, you will look at the different types of mixtures you can make and classify different mixtures as solutions, suspensions and colloids.

Solutions, suspensions and colloids

ICT opportunity

Watch a video about solutions, suspensions and colloids.

Write down an example of a solution, a suspension and a colloid.

solution: ....

suspension: ....

colloid: ....

Term 3 Unit 1 Mixtures

Making different mixtures

Try making these mixtures at home. Ask an adult first.

1. Mix each combination in a cup or a bowl.
2. Write down your observations of what each mixture looks like.
3. See if you can classify the mixtures as solutions, suspensions and colloids.

Mixture	Observations	Solution / suspension / colloid
Beat an egg white until it goes white and stiff.		
Mix oil and water.		
Whisk egg yolk with oil.		
Chalk and water		
Salt and water		
Flour and water		

Quick quiz

Are these statements **true** (T) or **false** (F)?

1. Colloids are transparent mixtures.   T   F
2. Suspensions are easily separated by filtration.   T   F
3. A mixture of oil and water is a solution.   T   F
4. Mixtures can be separated by different techniques.   T   F

Separating mixtures – filtration and sieving

> **Learning objectives**
> - Investigate the techniques of filtration and sieving to separate mixtures.

Make a note

A mixture of solids can be separated by sieving. A mixture of a liquid and a solid that doesn't dissolve can be separated by filtering or sieving depending on the size of the solid.

Complete the table to show if these mixtures can be separated by sieving or filtering.

Mixture	Sieving or filtering?
different particles in sand	
muddy water	
sand and water	
pasta and water	

Sieving

You will need:
- an old sieve
- a pair of gardening gloves
- garden soil.

Safety
Always wash your hands after handling soil.

Term 3 Unit 1 Mixtures

Use the sieve on some soil in your garden.

1. What happens to the larger soil particles and stones? Can you give a reason why?

   ....

   ....

2. What happens to the smaller soil particles? Can you give a reason why?

   ....

   ....

3. Look around your home. Can you write down any examples of sieving that happens in your home?

   Try to find three examples.

   a) ....

   b) ....

   c) ....

Planning a filtration practical

Task: You are going to plan to separate a mixture of sand and water.

Write down the apparatus you will use and a step-by-step method in the box on next page. Include a labelled diagram of the equipment.

Separating mixtures – filtration and sieving

Method	List of apparatus
1	
2	
3	
4	**Labelled diagram**
5	
6	
7	
8	

Quick quiz

1. Write down two different techniques you have used for separating mixtures.

   a) .... b) ....

2. Name a piece of equipment we need for sieving. ....
3. Name a piece of equipment we need for filtering. ....

Term 3 Unit 1 Mixtures

Separating mixtures – decanting and evaporation

> **Learning objectives**
> - Investigate the techniques of decanting and evaporation to separate mixtures.

You have come across the word 'evaporation' before this year.

Write down what you think evaporation means.

....

....

ICT opportunity

Now have a look at the word 'evaporation' in the dictionary or on the internet.

Write down what that definition says.

....

....

Are your ideas similar or the same as you found out?

In this lesson, you will learn about two different separation techniques, decanting and evaporation, and how these are used to separate mixtures.

Decanting

Make a note

Decanting is slowly pouring off the liquid so you separate it from the sediment (solid part).

Separating mixtures – decanting and evaporation

1. Mix some soil from your garden and water together in a glass. (Ask an adult first.)

Safety
Always wash your hands after handling soil.

2. Observe what your mixture looks like. Draw your observation in the box below.

3. Leave the mixture for two minutes.

4. Observe what the mixture looks like after two minutes. Draw your observation in the box below.

Appearance of mixture just after mixing	Appearance of mixture after two minutes

5. After two minutes, pour off as much of the water as you can. You should be left with soil. This technique is called 'decanting'.

How to grow a colourful crystal

Make a note

Evaporation is a technique that involves the removal of liquid from a solid by heating.

Term 3 Unit 1 Mixtures

1. Get some warm water and add a teaspoon of salt to it. Keep adding salt to it until no more salt will dissolve and you can see a few salt crystals at the bottom of the glass.

2. Add a few drops of food colouring to your mixture. You can use whatever colour you want your crystal to be.

3. Pour your solution into a shallow saucer and leave it on the windowsill to dry. Make sure that it won't be disturbed.

4. After a few days, the water from the mixture should evaporate leaving a coloured crystal.

5. Take a photo of your crystal or draw a picture of it in the box below.

My crystal:

Quick quiz

Are these statements **true** (**T**) or **false** (**F**)?

1. Decanting and evaporation can be used to separate liquids from solids.   T   F
2. We collect the liquid when we use the evaporation separating technique.   T   F
3. We can separate sand and water by decanting.   T   F

Separating using magnetism

> **Learning objectives**
> - Investigate using magnetism to separate mixtures.
> - Discover how magnetism is used to separate substances.

Make a note

Not all metals are magnetic.

ICT opportunity

Go online and research the names of metals that are magnetic.

Use the information you found and write three metals that are magnetic.

1 ....

2 ....

3 ....

In this lesson, you will learn about how magnetism is used to separate materials and some examples of where magnetism is used to separate materials.

Which is the best magnet?

A student separated iron filings from sand using different magnets.

She measured the mass of iron filings collected.

The results are shown in the table on the next page.

Term 3 Unit 1 Mixtures

Magnet	Mass of iron filings collected / g
A	12
B	11
C	15
D	9
E	11

1   Which is the best magnet? ....

2   Which two magnets picked up the same mass of iron filings?

   ....

3   What is the difference in mass between the iron filings collected by the best and worst magnet?

   ....

4   Describe how the student could make this a fair test.

   ....

   ....

   ....

Separating waste

ICT opportunity

Watch a video about separating recyclable materials.

Separating using magnetism

1. Describe how tin cans are separated from the rest of the waste.

   ....

   ....

2. Explain why we can't use the same magnets to separate aluminium cans.

   ....

3. How do aluminium cans get separated from the rest of the waste?

   ....

   ....

Quick quiz

Complete the table to show which separation techniques you would use to separate these mixtures.

You can choose from filtering, decanting, evaporation, magnetism and sieving.

Mixture	Separating technique
oil and water	
tin cans and aluminium cans	
salt and water	
different size soil particles	
soil and water	

Term 3 Unit 1 Mixtures

Investigating separating mixtures

Learning objectives
- Investigate how to separate a mixture of sand, salt and water.

Write down all the separation methods that you can think of that we have explored in this topic so far.

....

....

You will explore two of these in the main activity.

In this lesson, you will learn about how to use different separation techniques to separate a mixture of sand and salt water.

Separating sand and salt water

1 Read the equipment you will need and the method for this investigation.

You will need:
- safety goggles
- a beaker
- a glass stirring rod
- a funnel
- filter paper
- a conical flask
- an evaporating basin
- a Bunsen burner
- a heatproof mat
- a tripod and gauze
- mixture of sand and salt
- water.

Investigating separating mixtures

Method

1. Pour the sand and salt mixture into a beaker and add water so the beaker is a third full and stir.
2. Put the filter paper in the funnel and place the funnel in the conical flask.
3. Pour the mixture into the funnel.
4. Collect the filter paper and put it on the windowsill to dry.
5. Collect the liquid in the conical flask and pour it into the evaporating basin.
6. Heat the liquid in the evaporating basin until most of the liquid is gone.
7. Carefully remove the evaporating basin and leave it on the windowsill for a few days for the rest of the liquid to evaporate.

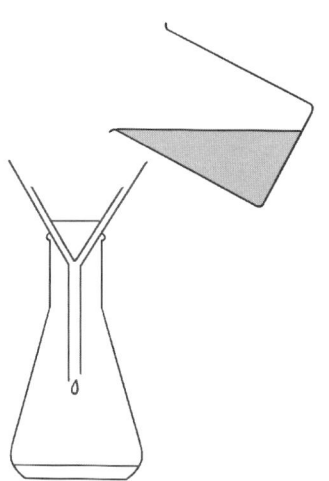

2. Answer the following questions about the investigation.

   a) Write down any safety measures the students should take for this investigation.

   ....

   ....

   ....

   b) What would we find on the filter paper at the end of this investigation?

   ....

Term 3 Unit 1 Mixtures

    c) What would we find in the evaporating basin at the end of this investigation?

    ....

    d) Which two separating techniques were used during this investigation?

    .... and ....

    e) Which substances went through the filter paper? Can you give a reason why these substances went through the filter paper?

    ....

    ....

    f) Suggest why the salt you collected might be contaminated with a little sand.

    ....

    ....

Quick quiz

Are these statements **true (T)** or **false (F)**?

1	Salt dissolves in water.	T	F
2	Filtration is a separation technique.	T	F
3	You can separate a solid and a liquid using evaporation and filtration.	T	F
4	You can separate two liquids using filtration.	T	F

Self-check

Learning objectives	😊	😐	☹
I can demonstrate that a mixture is made up of two or more substances.			
I can recognise that all mixtures can be separated.			
I can classify mixtures as solutions, suspensions and colloids.			
I can investigate the techniques of filtration and sieving to separate mixtures.			
I can investigate the techniques of decanting and evaporation to separate mixtures.			
I can investigate using magnetism to separate mixtures.			
I can discover how magnetism is used to separate substances.			
I can investigate how to separate a mixture of sand, salt and water.			

Term 3 Unit 1 Mixtures

Extension activity

Aim: To review the different separation techniques.

Task: To create a poster of the different ways that we can separate mixtures.

Include diagrams and colour to make your poster more interesting.

ICT opportunity

Research to find examples of the techniques being used.

Things to include:

1. What a mixture is.
2. Different types of mixtures.
3. The different types of separation techniques.
4. The equipment and methods you would need to carry out the different separation techniques.
5. Diagrams of the separation techniques.
6. Everyday examples of where these separating techniques are used.
7. Make notes on the next page and draw your poster on page 204.

Extension activity

Notes

Term 3 Unit 1 Mixtures

Poster

Practice test

Look at the table of different properties of different mixtures.

Mixture	Transparent / cloudy	Easily separated?
A	transparent	yes
B	transparent	no
C	cloudy	yes
D	cloudy	no
E	transparent	no
F	cloudy	no

1 Which mixtures are solutions? ....

2 Which mixtures are colloids? ....

3 What type of mixture is sugar and water? ....

4 Look at this list of substances.

   a) air   b) seawater   c) silver   d) oxygen   e) iced tea

Which substances can be separated? ....

5 Complete the sentence to describe a mixture.

Mixtures are substances made up of .... or .... other substances.

This diagram shows a separation technique.

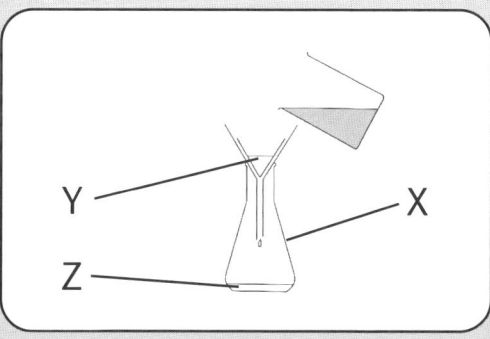

Term 3 Unit 1 Mixtures

6 Name the apparatus labelled X.

....

7 Name this separation technique.

....

8 Tick (✓) the boxes to show why the separation technique shown in the diagram works.

A	Some of the substances are larger than other substances in the mixture.	
B	Some of the substances are magnetic.	
C	Some of the substances are solid.	
D	The substances are both liquids.	
E	All the substances are small enough to fit through the filter paper.	

9 Which row shows the correct names for Y and Z in the diagram? Circle.

Row	Y	Z
A	solution	filtrate
B	residue	solution
C	filtrate	residue
D	residue	filtrate

A   B   C   D

This diagram shows a separation technique used to separate salt and water.

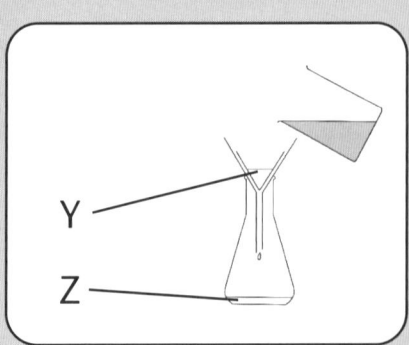

10 20 g of salt solution was evaporated and the mass of salt left was 5 g.

What is the mass of water that was evaporated? ....

11 Which is an explanation as to why the salt was left in the evaporation basin?

 a) Water has a very high boiling point.
 b) Water has a boiling point of 100°C.
 c) Salt has a boiling point of higher than 100°C.
 d) Water and salt have got different boiling points.

This table shows some different mixtures.

Mixture	Substances in the mixture
A	mud and water
B	sugar and water
C	iron and sand
D	peas and sand
E	oil and water

12 Which mixture(s) can be separated using decanting?

....

13 Which mixture(s) can be separated using magnetism?

....

14 Which mixture(s) can be separated using sieving?

....

15 Which of these metals are magnetic? Circle.

 a) steel   b) nickel   c) gold   d) aluminium

Term 3 Unit 2 Diet and drugs

Term 3 Unit 2 — Diet and drugs

Balanced diet

> **Learning objectives**
> - Identify what components are required for a balanced diet.

Make a note

Your body needs seven components of the diet to keep it healthy.

Unscramble the letters to give you the seven components of a balanced diet.

1 obhcaardyet ....

2 epnroit ....

3 taf ....

4 tewar ....

5 fireb ....

6 mivtisan ....

7 erlimsan ....

In this lesson, you will learn about balanced diets. You will look what components are needed for a balanced diet.

Term 3 Unit 2 Diet and drugs

Obesity

Learning objectives
- Identify the causes of obesity.
- Calculate BMI.
- Describe some of the health risks associated with obesity.

Make a note

When you eat too many fats and carbohydrates, you may become obese.

Circle the foods that might make you obese if you eat too much of them.

    carrots   mango   fries   patties   callaloo

In this lesson, you will look at what happens when we don't eat a balanced diet; in particular, what happens when you eat too much.

Calculating BMI

Make a note

One way to calculate obesity is using body mass index (BMI).
To calculate BMI, use this formula: BMI = mass (kg) / (height x height) (m).
A BMI of over 30 is classified as obese.

The table on the next page shows the height and mass of some people.

Calculate their BMI and classify them as obese or not. The first one has been done for you.

Obesity

Person	Height (m)	Height x height	Mass (kg)	BMI (mass / (height x height))	Obese? ✓ / ✗
A	1.6	2.56	55	55 ÷ 2.56 = 21.5	✗
B	1.8		88		
C	1.7		110		
D	1.6		88		

Finding out about obesity web activity

ICT opportunity

Go online and read information about obesity.

Use the information to answer the following questions.

1 What is obesity?

....

2 Name three causes of obesity.

a) ....

b) ....

c) ....

3 Name three health problems caused by obesity.

a) ....

b) ....

c) ....

Term 3 Unit 2 Diet and drugs

4 Name two ways to avoid obesity.

a) ....

b) ....

> **Quick quiz**
>
> Are these statements **true** (T) or **false** (F)?
> 1 Obesity causes lung cancer.   T   F
> 2 You can find out if a person is obese by using their BMI.   T   F
> 3 Too many vitamins in the diet causes obesity.   T   F
> 4 Exercise is one way to prevent obesity.   T   F

Diabetes

Learning objectives
- Describe the causes of and ways to prevent diabetes.

Make a note

There are two types of diabetes: type 1, which is genetic, and type 2, which develops later in life. Type 2 diabetes can be caused by having an unhealthy lifestyle.

Tick (✓) the factors which lead to an unhealthy lifestyle.

being overweight		eating lots of vegetables	
exercising regularly		being underweight	
smoking		having an active job	

Diabetes

In this lesson, you will look at what type 2 diabetes is. You will learn what the causes are and how to prevent it.

Diabetes

Make a note

Type 2 diabetes is a disease that you are more likely to get if you are obese.

ICT opportunity

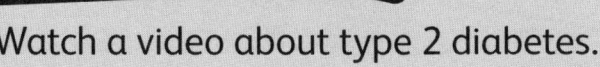

Watch a video about type 2 diabetes.

Use the information in the video to help you to answer the questions.

1 What is type 2 diabetes?

....

....

2 What are the symptoms of type 2 diabetes?

....

....

3 What can you do to reduce the risk of developing type 2 diabetes?

....

....

Term 3 Unit 2 Diet and drugs

Diabetes in Jamaica

This graph shows the percentages of men and women in the adult population that have type 2 diabetes. Use it to answer the questions.

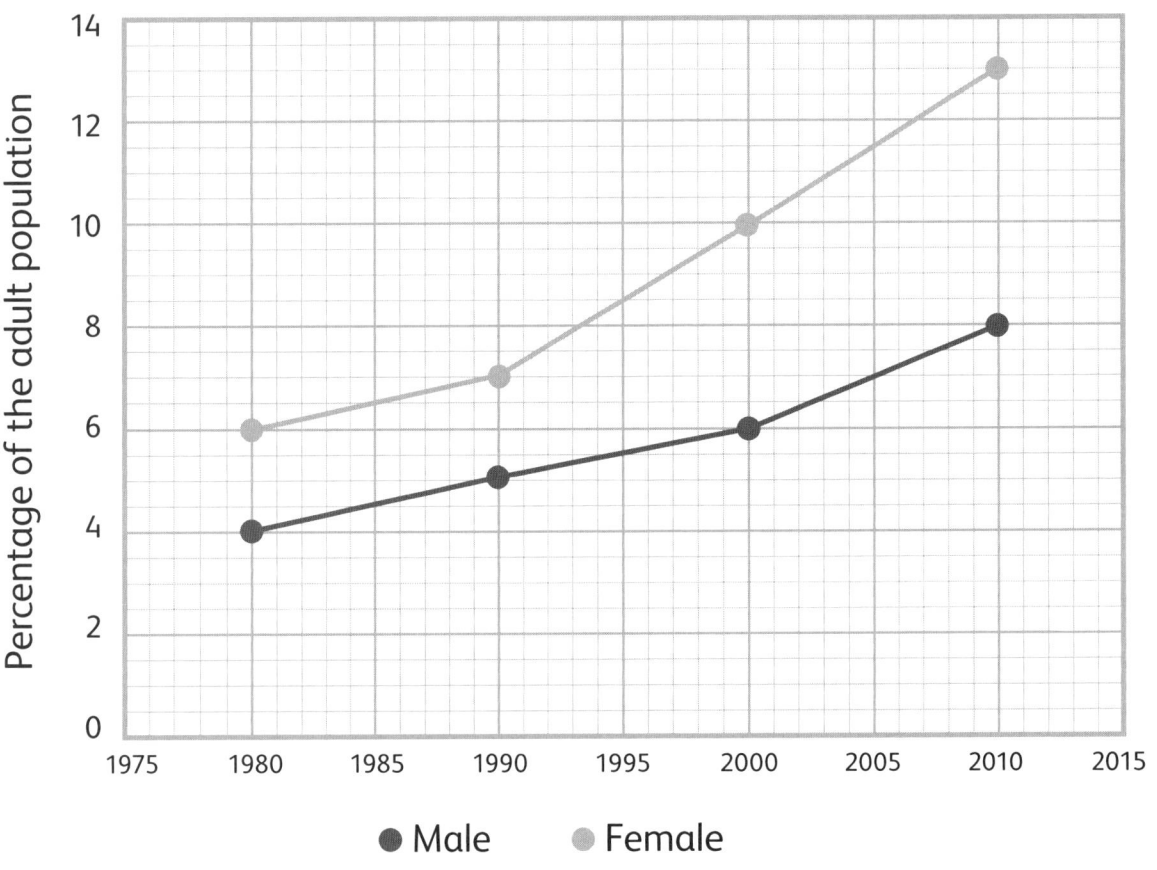

1 Describe the trend in the percentage of the population developing diabetes.

   ....

2 Calculate the difference in the percentage of women with diabetes between 1980 and 2010.

   ....

3 Estimate the percentage of men that had diabetes in 2015.

   ....

Malnutrition

> **Quick quiz**
>
> 1. Write down two risk factors of developing type 2 diabetes.
>
>   a) .... b) ....
>
> 2. Write down two things we can do to prevent developing type 2 diabetes.
>
>   a) .... b) ....

Malnutrition

> **Learning objectives**
> - Describe the causes of malnutrition.
> - Identify the different types of malnutrition.

Write down what you think 'malnutrition' means.

....

....

> **ICT opportunity**
>
> Now have a look at the word 'malnutrition' in the dictionary or on the internet.

Write down what that definition says.

....

....

Are your ideas similar or the same as you found out?

In this lesson, you will learn about what malnutrition is and the different types of malnutrition.

Malnutrition

> **Make a note**
>
> Malnutrition is when you do not get a balanced diet and you are deficient in one or more nutrients or energy.

Match the cause with the effect that we see in people.

Cause
too many carbohydrates in the diet
not enough energy in the diet
not getting one particular nutrient in your diet

Effect
underweight
deficiency disease
obesity

Marasmus and kwashiorkor

> **Make a note**
>
> Marasmus and kwashiorkor are two types of malnutrition.

> **ICT opportunity**
>
> Go online, research the two diseases, marasmus and kwashiorkor, and complete the information sheet on the next page.

Malnutrition

	Marasmus	**Kwashiorkor**
causes		
main symptoms		
treatment		

Quick quiz

Are these statements **true (T)** or **false (F)**?
1. Obesity is a type of malnutrition.   T   F
2. People suffering from kwashiorkor are deficient in fat.   T   F
3. Underweight people have a low BMI.   T   F

Term 3 Unit 2 Diet and drugs

Deficiency diseases

> **Learning objectives**
> - Describe the causes and effects of different deficiency diseases.

Make a note

Scurvy and rickets are two examples of deficiency diseases.

ICT opportunity

Go online and research the names of other deficiency diseases.

Use the information you found and write three other deficiency diseases.

1 ....

2 ....

3 ....

In this lesson, you will learn two different deficiency diseases (scurvy and rickets), their causes and their effects on the body.

Scurvy

Read this paragraph that tells the story of why Americans call British sailors 'limeys' and answer the following questions.

> Sailors in the British Navy had a rough life in the 1600s. After a few months at sea, a lot of them became unwell. They were tired and could barely walk. Their gums were swollen and their teeth started falling out. When you looked at their legs, they were swollen and purple from bruising. They had a disease called 'scurvy'.

Deficiency diseases

> A doctor called Dr Lind did an experiment and discovered their symptoms were due to vitamin C missing from their diets. Vitamin C is mainly found in fruits such as oranges, grapefruit, lemons, limes, strawberries and melons; it is also found in vegetables such as broccoli and bell peppers.
>
> When the sailors began their voyage, they had fresh fruit and vegetables on their ship. But fresh food goes off and, if the sailors were at sea for many months, they would not have fruit and vegetables for some of this time. This was a problem for the sailors, so the Royal Navy gave each sailor lime juice in their provisions if their journeys lasted longer than a month.
>
> This is how the English sailors became known as Limeys!

1 What are the symptoms of scurvy?

....

2 What causes scurvy?

....

3 Why were British sailors given lime juice in their provisions?

....

Rickets

ICT opportunity

Go online and find a picture of someone that has rickets.

1 Print it out or draw a picture in the box provided.

Term 3 Unit 2 Diet and drugs

2 Describe the effects of rickets on the body that you can see in the photograph.

....

....

....

....

3 What causes rickets?

....

4 What are the symptoms of rickets?

....

....

5 How can we prevent rickets?

....

....

Quick quiz

1 Scurvy is caused by a lack of which vitamin? ....

2 Bowed legs is a symptom of which deficiency disease? ....

3 A person has scurvy. What food should they add to their diet? ....

4 Name one way to prevent rickets that doesn't include diet. ....

Self-check

Learning objectives	☺	😐	☹
I can identify what components are required for a balanced diet.			
I can identify the causes of obesity.			
I can calculate BMI.			
I can describe some of the health risks associated with obesity.			
I can describe the causes and ways to prevent diabetes.			
I can describe the causes of malnutrition.			
I can identify the different types of malnutrition.			
I can describe the causes and effects of different deficiency diseases.			

Term 3 Unit 2 Diet and drugs

Legal and illegal drugs

> **Learning objectives**
> - Classify some drugs as legal or illegal.
> - Understand that some legal drugs can be harmful if misused.

Classify the following drugs into legal or illegal.

1. cannabis ....
2. tobacco ....
3. alcohol ....
4. paracetamol ....
5. heroin ....
6. cocaine ....

In this lesson, you will learn about different types of drugs. You will look at the difference between legal and illegal drugs and be able to explain that legal drugs can be harmful.

Legal drugs

Make a note

Drugs are substances that when taken affect your body or your mind.
Legal drugs can be classified as medicinal or recreational.

1 Write down what you think the words 'medicinal' and 'recreational' mean.

medicinal: ....

recreational: ....

Legal and illegal drugs

> **Make a note**
>
> Some legal recreational drugs are controlled by the government and have age restrictions. Legal drugs are harmful if they are misused.

> **ICT opportunity**
>
> Go online and research the legal age for drinking alcohol and smoking in the following countries.

2 Complete the table.

Country	Legal age for drinking alcohol	Legal age for smoking
Jamaica		
UK		
USA		
France		
Japan		

More about legal and illegal drugs

Read about the effects of three different drugs on the body.

Drug A	Drug B	Drug C
strong painkiller	loss of inhibition	addiction
addiction	slurred speech	coughing
overdose can lead to respiratory failure and death	vomiting	shortness of breath
	coma/death	increased heart rate

Term 3 Unit 2 Diet and drugs

1. Can you put the drugs in order of most dangerous to least dangerous?

   ....

2. All of these drugs are legal. Which do you think is the medicinal drug?

   ....

3. One of the drugs is alcohol, one is morphine (a medicinal drug), and one is tobacco.

   Which drug do you think is which?

   Drug A: ....

   Drug B: ....

   Drug C: ....

4. Why do you think that some drugs are legal even though they are harmful? Write down your ideas.

   ....

   ....

Quick quiz

Are these statements **true** (T) or **false** (F)?

1 Heroin is a legal drug.	T	F
2 Legal drugs are not harmful.	T	F
3 Alcohol is a legal drug.	T	F
4 It is legal to possess cocaine but not sell it.	T	F

More about illegal drugs

> **Learning objectives**
> - Research an illegal drug.
> - Describe the effects of addiction and withdrawal.

Write down the names of as many illegal drugs as you can think of. How many did you name?

....

....

....

In this lesson, you will learn more about illegal drugs. You will research an illegal drug and find out about what happens if people become addicted to drugs.

Illegal drugs

Make a note

It is against the law for people to possess, take or distribute illegal drugs.

1. You are going to research an illegal drug. You can choose cannabis, heroin or cocaine.

ICT opportunity

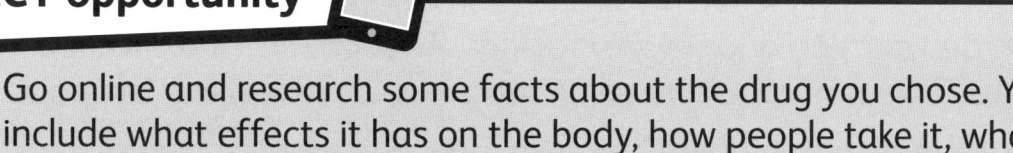

Go online and research some facts about the drug you chose. You can include what effects it has on the body, how people take it, what are the consequences if the police find you have the drug.

Term 3 Unit 2 Diet and drugs

2 Create a factsheet about your chosen drug in the box below. Try to find out at least six facts.

Addiction and withdrawal

This is a diary of a person who tried to stop using heroin.

> At first, I took drugs socially and then it became a habit. I was using drugs every day and I would get very angry if I didn't get any. My lowest point was when I stole money from my mother to get the drugs I needed to feed my addiction.
>
> I decided I needed to stop taking drugs. I checked myself into a motel room away from everyone I knew. The first day, the withdrawal symptoms began. I had runny eyes and nose, diarrhoea, dehydration, complete lack of appetite, I felt sick and my whole body was in pain.
>
> The other problem is that I couldn't sleep. I was anxious that I was never going to get better. I didn't leave my motel room for five days. I was suffering from dehydration and severe cravings. As soon as I had checked out of my motel, I went out to score some drugs just to take away the pain.
>
> After several attempts at rehabilitation, I am now one month drug-free. Every day is a struggle.

More about illegal drugs

Read the diary entry to answer the following questions.

1 What do we mean by the word 'addiction'?

....

2 Name some of the withdrawal symptoms the person experienced.

....

....

3 Why did the person take heroin again?

....

Quick quiz

1 Name three illegal drugs.

a) .... b) .... c) ....

2 Name three withdrawal symptoms.

a) .... b) .... c) ....

3 Describe what 'addiction' means.

....

Medicinal drugs

> **Learning objectives**
> - Describe the difference between prescription and over-the-counter drugs.
> - Examine the information on a medicine packet.

Classify these drugs into medicinal or not medicinal.

1 heroin   ....

2 paracetamol   ....

3 penicillin   ....

4 alcohol   ....

5 ibuprofen   ....

6 tobacco   ....

In this lesson, you will learn about the difference between prescription and over-the-counter drugs. You will also extract some information from a medicine packet.

Prescription drugs and over-the-counter drugs

Make a note

Over-the-counter drugs can be bought without a prescription at a pharmacy. Prescription drugs requires a note from the doctor called a 'prescription' to get them.

ICT opportunity

Use the internet to find out the names of three prescription drugs and three over-the-counter drugs.

Medicinal drugs

1 Complete the table.

Prescription drugs	Over-the-counter drugs
a)	a)
b)	b)
c)	c)

2 Why you think that some medicines are available over-the-counter and some have to be prescribed? Write down your ideas.

....

....

Medicine packets

Read the packaging on this medicine packet and answer the following questions.

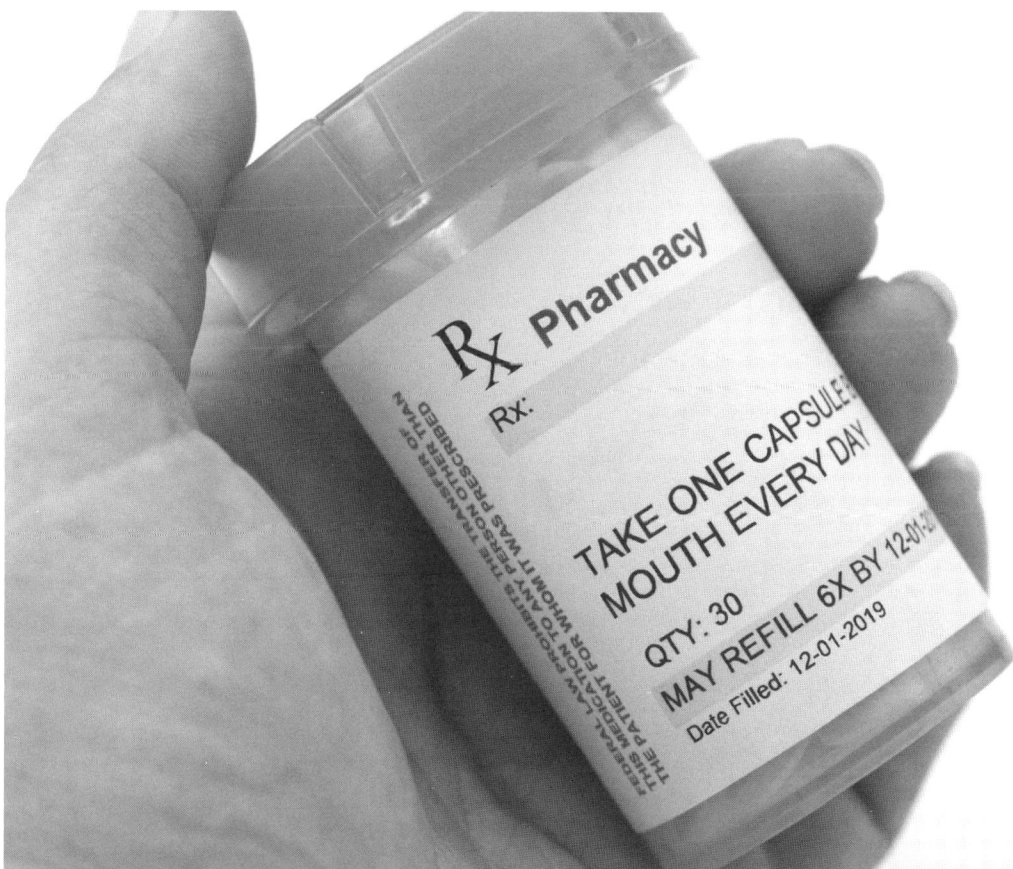

Term 3 Unit 2 Diet and drugs

1 What is the recommended dose?

....

2 How many doses are in this packet?

....

3 Is this bottle refillable?

....

4 When it was first produced?

....

5 How should this medicine be administered?

....

Quick quiz

Are these statements **true** (**T**) or **false** (**F**)?

1	Medicinal drugs are not harmful.	T F
2	All medicinal drugs need a prescription.	T F
3	Over-the-counter drugs are generally less harmful than prescription medicines.	T F

Alcohol

> **Learning objectives**
> - Describe the short-term and long-terms effects of alcohol on the body.

Make a note

There are many different types of alcoholic drinks that come in different strengths.

Write down as many different types of alcoholic drinks as you can think of in the space below.

....

....

In this lesson, you will learn about the short-term and long-term effects of alcohol on the body.

The effects of alcohol on the body

Make a note

Alcohol is a legal drink. If alcohol is misused, it can have severe negative effects on the body.

ICT opportunity

Watch a video about the effects of alcohol on the body.

Term 3 Unit 2 Diet and drugs

1 Describe three short-term effects of alcohol on the body.

   a) ....

   b) ....

   c) ....

2 Name two organs that are affected by long-term alcohol consumption.

   a) ....

   b) ....

3 Name three types of cancer you can get from drinking alcohol.

   a) ....

   b) ....

   c) ....

4 Is alcohol addictive? ....

5 Can you name one other drug that is addictive? ....

Type of alcohol drunk in Jamaica

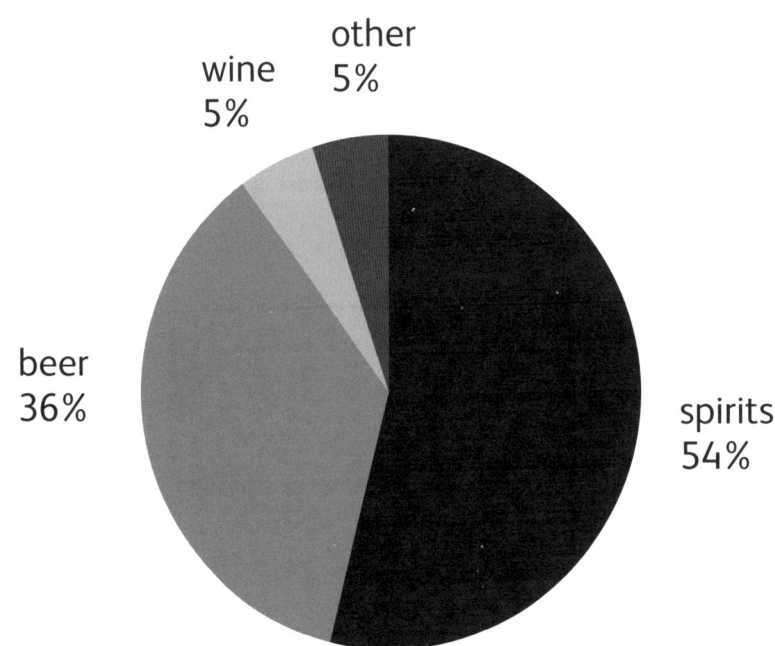

beer 36%
wine 5%
other 5%
spirits 54%

Alcohol

The pie chart on the previous page shows the different types of alcohol drunk in Jamaica.

1. What is the most popular type of alcoholic drink in Jamaica?

   ....

2. What percentage of people drinking alcohol choose to drink beer?

   ....

3. What is the total percentage of people that drink beer and wine?

   ....

Alcohol warning

Imagine you are a doctor and you have to warn some teenagers about the effects of alcohol before they go to a party. What advice would you give them?

Write it in the box below.

```
................................................................................................
................................................................................................
................................................................................................
................................................................................................
................................................................................................
```

Quick quiz

Are these statements **true (T)** or **false (F)**?

1. Alcohol can cause death.   T   F
2. Alcohol is an illegal drug.   T   F
3. Alcohol damages the brain and liver. T   F

Tobacco

Learning objectives
- Describe the effects of tobacco on the body.

ICT opportunity

Watch a cartoon talking about smoking.

What message do you think this cartoon is trying to tell the viewers? Write down your ideas.

....

....

In this lesson, you will learn about tobacco and smoking. You will look at the effects of smoking and the link between smoking and lung cancer.

What does tobacco do to your body?

ICT opportunity

Go online and read information about smoking.

Answer the following questions.

1 Name some short-term effects of smoking.

....

....

2 What is the addictive drug in tobacco? ....

3 What serious disease does smoking cause? ....

4 A parent is trying to give up smoking. Write a list of reasons why they should give up.

....

....

....

....

....

Smoking and lung cancer

This graph shows the link between smoking and lung cancer.

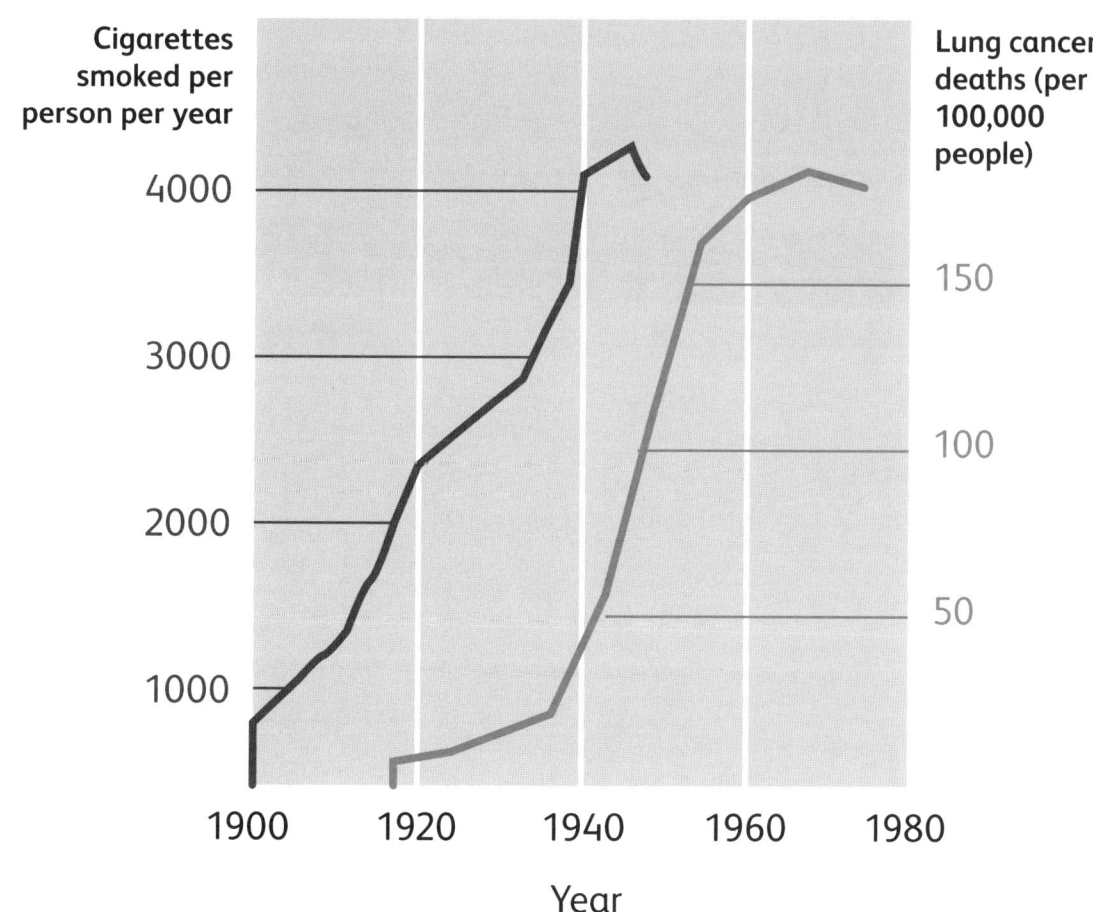

Term 3 Unit 2 Diet and drugs

1 What has happened to the number of cigarettes smoked per year?

....

2 Approximately what year was the most cigarettes smoked?

....

3 Approximately what year was the most cases of lung cancer?

....

4 Suggest why the peak of lung cancer was after the peak of cigarettes smoked.

....

....

Quick quiz

Circle the correct words.

1. The addictive drug in tobacco is **adrenaline** / **cocaine** / **nicotine**.
2. Tobacco affects your **liver** / **lungs** / **stomach**.
3. Smokers often develop coughs and colds because their **immune** / **reproductive** / **digestive** system is affected.

Lifestyle and health

> **Learning objectives**
> - Describe what a healthy lifestyle consists of.
> - Evaluate different lifestyles and describe ways to improve them.

Tick (✓) the factors that would make your lifestyle healthy.

being obese		regular exercise	
eating fruit and vegetables		drinking alcohol	
smoking		being underweight	

In this lesson, you will learn about what a healthy lifestyle is. You will look at how people can make improvements to their lifestyle to make them healthier.

A healthy lifestyle

Make a note

A healthy lifestyle is a way of living that helps you to keep your mind and body functioning well. Both your mental health and your physical health are important.

1. Think about what makes a healthy lifestyle. Write down your ideas.

   ....

   ....

   ....

Term 3 Unit 2 Diet and drugs

2 Think about your lifestyle. Do you think it is healthy?

a) Write down the parts of your lifestyle that help to keep you healthy.

....

b) Write down the parts of your lifestyle that are not so healthy.

....

c) Write down one improvement to your lifestyle that you could make that would make you healthier.

....

Different lifestyles

Imagine you are a doctor. Read the different patient files.

Patient A		Patient B		Patient C	
age:	22	age:	66	age:	44
sex:	male	sex:	female	sex:	female
occupation:	athlete	occupation:	office worker	occupation:	business owner
hobbies:	running	hobbies:	watching films	hobbies:	swimming
diet:	lots of fruit and vegetables	diet:	high fat diet	diet:	vegetarian
BMI:	overweight	BMI:	obese	BMI:	underweight
other:	non-smoker	other:	smoker	other:	occasional drug and tobacco use

What advice would you give to each patient to improve their lifestyle?

Patient A: ....

....

Patient B: ....

....

Patient C: ....

....

Quick quiz

Are these statements **true** (**T**) or **false** (**F**)?

1	Being underweight is healthy.	T F
2	A healthy diet is part of a healthy lifestyle.	T F
3	Reducing stress can improve your lifestyle.	T F
4	Smoking is OK if you exercise.	T F

Term 3 Unit 2 Diet and drugs

Self-check

Learning objectives	☺	😐	☹
I can classify some drugs as legal or illegal.			
I can understand that some legal drugs can be harmful if misused.			
I can research an illegal drug.			
I can describe the effects of addiction and withdrawal.			
I can describe the difference between prescription and over-the-counter drugs.			
I can examine the information on a medicine packet.			
I can describe the short-term and long-terms effects of alcohol on the body.			
I can describe the effects of tobacco on the body.			
I can describe what a healthy lifestyle consists of.			
I can evaluate different lifestyles and describe ways to improve them.			

Extension activity

Aim: To review the components of a healthy lifestyle.

Task: To create an information leaflet aimed at young adults to encourage them adopt a healthy lifestyle.

Include scientific facts to provide evidence in your leaflet.

ICT opportunity

Go online and research information to include in your leaflet.

Things to include:

1. What a healthy lifestyle consists of.
2. What a balanced diet is and why you need a balanced diet.
3. What things happen to our bodies if we don't have a balanced diet.
4. What drugs are.
5. What smoking and alcohol do to your body.
6. What taking illegal drugs does to your body.
7. Make notes on the next page and draw your leaflet on page 243.

Term 3 Unit 2 Diet and drugs

Notes

Extension activity

Leaflet

Term 3 Unit 2 Diet and drugs

Practice test

This diagram shows a label from a food packet.

Nutrition Facts
Per 1 cup (55 g)

Amount	% Daily Value
Calories 220	
Fat 2 g	3 %
Saturated 0 g + Trans 0 g	0 %
Cholesterol 0 mg	
Sodium 270 mg	11 %
Carbohydrate 44 g	15 %
Fibre 8 g	32 %
Sugars 16 g	
Protein 6 g	
Vitamin A 0 % Vitamin C 0 %	
Calcium 4 % Iron 40 %	

1. Which important component of the diet is this food stuff missing?

   ....

2. If a person only ate this food stuff which deficiency disease would they be at risk from? Circle.

   a) obesity

   b) scurvy

   c) rickets

   d) marasmus

3. How many cups of this food stuff would a person need to get their daily percentage of iron?

   ....

Look at the shopping list.

Shopping list
tuna fish
milk
yams
sweet peppers
grapefruit

4. Which food from the list should a person eat if they were most interested in building muscles?

   ....

5. Which food from the list should a person eat to prevent rickets?

   ....

244

This table shows the BMI of some people.

Person	A	B	C	D
BMI	35	21	24	40

6 Which of these people (A–D) are most at risk of developing diabetes?

....

This graph shows the percentage of people that have smoked, ex-smokers and people that have never smoked in a population.

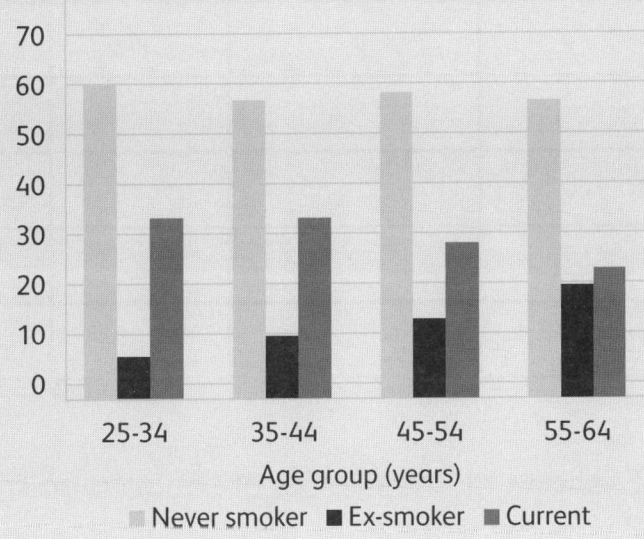

7 Tick (✓) the boxes to show which conclusions we can read from the graph.

A	The 35-to-44-year-old age group has the highest number of current smokers.	
B	The 55-to-64-year-old age group has the fewest current smokers.	
C	The 55-to-64-year-old age group has the highest number of ex-smokers.	
D	There are more current smokers in the 35-to-44-year-old age group than the 45-to-54-year-old age group.	
E	The largest group in each age range is the people who have never smoked.	

Term 3 Unit 2 Diet and drugs

8 Which statement explains why there are more people in the 25-to-34-year-old age group that have never smoked than the 55-to-64-year-old age group? Circle.

a) The link between lung cancer and smoking is now well-known.

b) The legal age of smoking is over 30.

c) Older people are more careful with their health.

d) More younger people develop lung cancer.

This table shows some short-term and long-term effects of different drug.

	Lung cancer	Liver damage	Loss of coordination	Memory loss	Shortness of breath	Coma
A		✓	✓	✓		✓
B	✓					
C		✓				
D			✓	✓		
E	✓				✓	

9 Which row (A–E) shows all the effects of tobacco on the body?

....

10 Which row (A–E) shows the short-term effects of alcohol abuse only?

....

Patient 1	Patient 2	Patient 3
50 years old	24 years old	35 years old
heavy smoker	high fat diet	fitness instructor
high fat diet	obese	balanced diet
obese	excessive alcohol consumption	healthy BMI

Practice test

Patient 4	Patient 5	Patient 6
56 years old	19 years old	25 years old
high-stress job	heavy smoker	heavy smoker
heavy smoker	drug user	normal BMI
high fat diet	underweight	office job
normal BMI	alcohol consumption in moderation	non-drinker

11 Which two patients are most at risk from developing diabetes?

    .... and ....

12 Which patient has the healthiest lifestyle? ....

13 Which advice should be given to Patient 6 to improve their lifestyle? You can circle more than one.
   a) eat less calories
   b) stop drinking alcohol
   c) stop using illegal drugs
   d) stop smoking
   e) take regular exercise

14 Why prescription drugs have to be prescribed by a doctor? You can circle more than one.
   a) Prescription drugs are illegal.
   b) Prescription drugs are used for recreational purposes only.
   c) Prescription drugs are medicines.
   d) Prescription drugs can be dangerous if taken incorrectly.

15 Complete the sentences to describe the meaning of the words 'addiction' and 'withdrawal'.
   a) Addiction is when your body has built up a tolerance to the drug and you don't feel .... unless you take it.
   b) Withdrawal symptoms are what you feel when you .... taking the drug.

Acknowledgements

Photo acknowledgements

The publishers are grateful to the following for permission to reproduce copyright photographs and material:

T = Top, B = Below, L = Left, R = Right, C = Centre, B/G = Background

p.5: © Daniel Prudek/stock.adobe.com; p.6: (T) © Richard Carey/stock.adobe.com, (C) © Alexmar/stock.adobe.com; p.10: (camel) © Leo Lintang/stock.adobe.com, (polar bear) © Zahi/stock.adobe.com, (shark) © Le Bouil Baptiste/stock.adobe.com, (kapok tree) © oldmn/stock.adobe.com; p.17: © Rixie/stock.adobe.com; p.23: © jon manjeot/stock.adobe.com; p.25: © vivanvu/Shutterstock.com; p.28: © JAH/stock.adobe.com; p.46: © Apisit/stock.adobe.com; p.59: © ktsdesign/stock.adobe.com; p.80: © Lakshmiprasad/stock.adobe.com; p.84: © trodler1/stock.adobe.com; p.97: © Michal Sanca/Shutterstock.com; p.101: © anon_tae/Shutterstock.com; p.104: (TL) © Tsvetina/stock.adobe.com, (TR) © sudowoodo/stock.adobe.com, (B) © Tsvetina/stock.adobe.com; p.110: (1) © Goffkein/stock.adobe.com, (2) © voyata/stock.adobe.com, (3) © max3d007/stock.adobe.com, (4) © magdal3na/stock.adobe.com, (5) © Yevhenii/stock.adobe.com, (6) © Roman Motizov/stock.adobe.com; p.119: (sugar) © Cozine/stock.adobe.com, (coffee) © BillionPhotos.com/stock.adobe.com, (water) © alter_photo/stock.adobe.com; p.120: (balloons) © New Africa/stock.adobe.com, (rocks) © pamela_d_mcadams/stock.adobe.com, (ice) © Людмила Короткова/stock.adobe.com, (steam) © kazoka303030/stock.adobe.com, (wool) © MaciejBledowski/stock.adobe.com, (cola) © Alex/stock.adobe.com; p.142: © GraphicsRF/stock.adobe.com; p.144: (a, b, c, e, f) © GraphicsRF/stock.adobe.com, (d) © VectorMine/stock.adobe.com; p.146: © beeny1983/stock.adobe.com; p.147: © GraphicsRF/stock.adobe.com; p.149: © GraphicsRF/stock.adobe.com; p.151: © VectorMine/stock.adobe.com; p.154: © GraphicsRF/stock.adobe.com; p.160: © GraphicsRF/stock.adobe.com; p.163: (T) © GraphicsRF/stock.adobe.com, (B) © paveu/stock.adobe.com; p.170: © LuckySoul/stock.adobe.com; p.182: © toricheks/stock.adobe.com; p.183: © Alexandre/stock.adobe.com; p.208: © PhotoIris2021/stock.adobe.com; p.229: © Sherry Young/stock.adobe.com.